Engineering
Instrumentation
and Control

Engineering Instrumentation and Control

W. Bolton

Technician Education Council Senior Adviser

Butterworths
London Boston
Sydney – Wellington – Durban – Toronto

THE BUTTERWORTH GROUP

UNITED KINGDOM
Butterworth & Co. (Publishers) Ltd
London: 88 Kingsway, WC2B 6AB

AUSTRALIA
Butterworths Pty Ltd
Sydney: 586 Pacific Highway, Chatswood, NSW 2067
Also at Melbourne, Brisbane, Adelaide and Perth

CANADA
Butterworth & Co. (Canada) Ltd
Toronto: 2265 Midland Avenue, Scarborough, Ontario M1P 4S1

NEW ZEALAND
Butterworths of New Zealand Ltd
Wellington: T & W Young Building, 77–85 Customhouse Quay, 1, CPO Box 472

SOUTH AFRICA
Butterworth & Co. (South Africa) (Pty) Ltd
Durban: 152–154 Gale Street

USA
Butterworth Publishers, Inc.
Boston: 10 Tower Office Park, Woburn, Mass. 01801

First published 1980

© Butterworth & Co. (Publishers) Ltd, 1980

British Library Cataloguing in Publication Data

Bolton, William, b. 1933
    Engineering instrumentation and control
    – (Technician Education Council. TEC technician series).
    1. Engineering instruments
    2. Measuring instruments 3. Control theory
    I. Title II. Series
    620'.0044 TA165 79–41524

    ISBN 0-408-00462-2

Typeset by Reproduction Drawings Ltd, Sutton, Surrey
Printed in England by Page Bros Ltd., Norwich, Norfolk

Acknowledgements

Thanks are due to the following manufacturers for permitting reproductions of items and pages from their catalogues and for their assistance in the preparation of the book.

Griffin & George Ltd, P.O. Box 14, Wembley HAO 1 HJ.

A.P.V.-Osborne Craig Ltd, Glenburn Road, College Milton North, East Kilbride, Glasgow G74 5BJ.

Vibro-meter Ltd, Newby Road, Hazel Grove, Stockport, Cheshire SK7 5EE.

J. J. Lloyd Instruments Ltd, Brook Avenue, Warsash, Southampton SO3 6HP.

Philip Harris Ltd, Lynn Lane, Shenstone, Staffordshire WS14 OEE.

C. Stevens & Son (Weighing Machines) Ltd, 287/289 Goswell Road, London EC1V 7LD.

Sacol Controls Ltd, Commercial Road, Totton, Southampton SO4 3ZQ.

Tektronix UK Ltd, P.O. Box 69, Coldharbour Lane, Harpenden, Herts, AL5 4UP.

Davy Instruments Ltd, Darnall Works, Sheffield, S9 4FA.

Penny & Giles Conductive Plastics Ltd, Newbridge Road Industrial Estate, Pontllanfraith, Blackwood, Gwent NP2 2 YD.

Appendix B is reproduced by permission of the Editor of *Engineering*.

Preface

This book has been written with the following aims:

1. To provide a basic knowledge of the different types of instrumentation encountered in engineering.
2. To permit selection of the appropriate measurement system for a specific purpose.
3. To assist understanding of the jargon of manufacturers' catalogues and data sheets as an aid to intelligent selection of a measurement system for a particular purpose.
4. To introduce the basic concepts of control systems.
5. To provide a basic understanding of the response of systems to external forces.

The book covers the unit 'Engineering Instrumentation and Control' (TEC U77/422) of the Technician Education Council. This unit is considered to be part of an essential core of higher certificates and higher diplomas in mechanical and production engineering. Covering all aspects of the unit and extending the measuring systems into additional areas, the book is also likely to be of use in other engineering courses; in fact in any course where a basic knowledge of engineering instrumentation and control is required to the depth indicated by the above aims.

Contents

1 Systems

Aims: At the end of this chapter you should be able to:
Recognise the basic elements of a measuring system.
Explain the common terms applied to measuring systems.
Interpret catalogue specifications of measuring systems.

This chapter is intended as a reference for terms that are encountered in later chapters.

MEASURING SYSTEMS

Figure 1.1 shows a Bourdon tube pressure gauge—a system for the measurement of pressure. The Bourdon tube is an example of a *transducer*, in which an increase in pressure causes the tube to straighten a

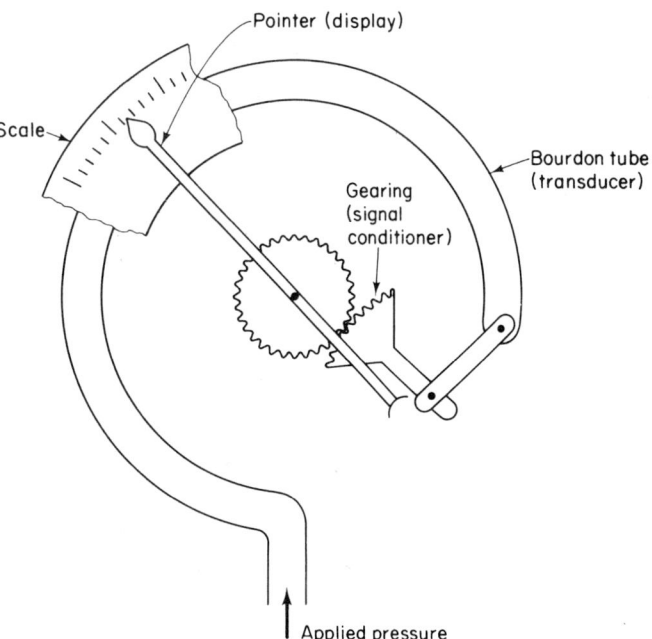

Figure 1.1 Bourdon tube pressure gauge—a measurement system

little. The input of pressure is thus changed into mechanical displacement of the tube. In this case the transducer changes information about pressure to information in the form of a mechanical displacement. Transducers change information from one form to another.

$$\text{pressure information} \xrightarrow{\text{Transducer}} \text{displacement information}$$

The displacement of the tube is comparatively small and needs to be made larger for display and reading; this is done by gearing, which performs the function of a *signal conditioner*. In general, a signal conditioner converts the signal from the transducer into a form which can be displayed. Without the gearing the output from the Bourdon tube would be too small to be read easily.

tube displacement $\xrightarrow{\text{Signal conditioner}}$ bigger displacement of centre gear

The movement of the gear wheel causes a pointer to move across a scale and the pointer and the scale constitute what is called the *display element*.

A mercury-in-glass thermometer is a measuring system. When the temperature increases the volume of the mercury increases, so that here the mercury is the transducer. Information about the temperature (invisible) is converted into information in the form of a volume change (visible).

temperature information $\xrightarrow{\text{Transducer}}$ volume information

The volume change is quite small but is converted into a noticeable change in length by the mercury, being contained in a small diameter capillary tube.

volume change $\xrightarrow{\text{Signal conditioner}}$ length change

The display which enables a measurement to be made is given by the mercury expanding or contracting in the capillary tube which has a graduated scale beside it.

length change $\xrightarrow{\text{Display}}$ position of meniscus of mercury against a scale

In general, measurement systems can be represented as having three elements:

(1) A detecting element called a transducer which produces a signal related to the quantity being detected.
(2) An element called a signal conditioner which converts the signal from the transducer into a form which can be displayed.
(3) A display or recording element which enables the signal to be read.

Figure 1.2 is a representation of this generalised measurement system. The input to the transducer produces an output signal which is then converted into a suitable form for display.

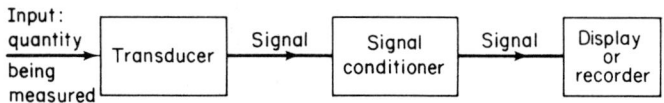

Figure 1.2 A measuring system

THE PERFORMANCE OF MEASURING SYSTEMS

The advertisement for a thermometer reads as follows:

Range and subdivision °C
  −0.5 to +40.5 × 0.1

Maximum error °C
      0.2

This indicates that the thermometer, a mercury-in-glass type, can be used for temperature measurement between −0.5°C and +40.5°C and has a scale which is subdivided into 0.1°C intervals. If the meniscus of the mercury in the thermometer were at the 30.1°C mark on the scale then the manufacturer would claim only that the actual temperature be within 0.2°C of that reading. The actual temperature could thus lie between 29.9°C (i.e. 30.1 − 0.2) and 30.3°C (i.e. 30.1 + 0.2). This is the maximum *error* possible under the conditions specified for the use of the thermometer. In this case the specification is for the thermometer to be totally immersed in the environment for which the temperature is being measured. Thus if the thermometer is being used to measure the temperature of a liquid then it should be totally immersed in the liquid. Only under these conditions is the error a maximum of 0.2°C.

    The *accuracy* of a measuring system is the closeness with which the readings given by the system approach the true values of the quantities being measured. The temperature measuring system above is more accurate than one with a quoted maximum error of 0.4°C. The greater the maximum error quoted the lower the accuracy.

    Another thermometer is quoted as having the following specification:

Range and subdivision °C
  −20 to +60 × 0.2
Total immersion.

Maximum error °C
      0.3

Such a thermometer can be used for the measurement of temperatures between −20 and +60°C and has a scale which is subdivided into 0.2°C. When the thermometer is used totally immersed the maximum error is 0.3°C, which means that any reading on the scale indicates a tempera-ture within a region bounded by plus or minus 0.3°C of the indicated value. Thus if the thermometer gives a reading of 25.6°C the actual temperature is between 25.3°C and 25.9°C.

    The following is part of the specification of a balance:

Capacity 250 g
Sensitive to 1 mg

This means that the instrument can be used for measurements up to 250 g and mass changes of 1 mg can be detected. It does not state the accuracy of the instrument. This depends on both the weights used with the balance and the changes that can be detected by the balance. Masses with an accuracy of least 1 mg should be used if the full capa-bilities of the instrument are to be used.

    The following is taken from a catalogue:

Tolerances for masses

Denomination	10	20	50	100 mg	10	20	50	100 g
Tolerance ± (mg)	0.02	0.02	0.02	0.05	0.05	0.1	0.25	0.5

The term *tolerance* indicates the maximum error. The weights are more than accurate enough for the balance described above.

The following is part of the specification of a hydrometer, an instrument used for the measurement of the density of liquids:

Range 600 to 650 kg/m³
Scale subdivision 1kg/m³
Tolerance ± 0.6 kg/m³

Such an instrument can be used for the measurement of densities between 600 and 650 kg/m³. The scale of the instrument is divided into intervals of 1 kg/m³. The accuracy is ± 0.6 kg/m³ and thus it is possible to measure densities to an accuracy slightly greater than that indicated by the scale of the instrument.

Accuracy can be specified in terms of the *percentage of the true value*, i.e. in the form

$$\text{error} = \frac{\text{measured value} - \text{true value}}{\text{true value}} \times 100 \text{ per cent}$$

Another way of specifying accuracy is in terms of *percentage of full-scale deflection* and is often used where the system gives a result on a scale.

$$\text{error} = \frac{\text{measured value} - \text{true value}}{\text{maximum scale value}} \times 100 \text{ per cent}$$

The specification for a moving iron ammeter with two current ranges reads as follows:

Ranges 0 – 3 and 0 – 30 A
Accuracy ± 1% (of the full-scale deflection)

This means that if the ammeter is connected so that the 0 to 3 A range is used then the accuracy is ±1% of 3 A and so ± 0.03 A. With the 0 to 30 A range the accuracy for a current of 30 A is ±0.3 A. Accuracy, or error, is often expressed for meters as a percentage of the full measuring range of a system.

If a thermometer is used to measure the temperature of an object at, say, 30.0°C then every time the thermometer is used to measure that temperature a concern would be that the same temperature reading was indicated by the thermometer. For the same input to the measuring system the concern is whether the same output is always obtained. The term *repeatability* is used to express the ability of a system to display the same output for a series of the applications of the same input signal, the time intervals between the applications being relatively short. The term *stability* is used to express the ability of a system to display the same output for a series of applications of the same input signal when the time intervals between the applications is long.

If a thermometer is placed in a constant temperature enclosure at, say, 30.0°C and left there then the concern is whether the thermometer will indicate a temperature of 30.0°C and whether it will continue to indicate this temperature however long it is left in the enclosure. The term *constancy* is used to describe the ability of a system to hold a constant display reading in response to a constant input.

The term *reproducibility* is used to describe the ability of a system to display a reading for a given input when that input is either constantly applied to the system or presented to the system on a series of occasions. The terms repeatability, stability and constancy are just ways of expressing the reproducibility of a system under different input conditions.

A galvanometer has a sensitivity specified of 17 mm/μA. This means that for a 1 μA input the display, in this case a light spot moving across a scale, shows a movement of an index of 17 mm. Another galvanometer is quoted as having a sensitivity of 43 mm/μA. This galvanometer gives a deflection of an index of 43 mm for every 1 μA input. This is a greater sensitivity than the first galvanometer, the index moving more for the same input.

The *sensitivity* of a measuring system or part of such a system is defined as follows:

$$\text{sensitivity} = \frac{\text{movement of the index}}{\text{change in the quantity producing the movement}}$$

In the case of the galvanometer the sensitivity is expressed in millimetres of movement of a light spot across a scale per microampere of input. As the instrument can be read to 1 mm this means the sensitivity is 17 mm/μA and that a change in current of 1/17 microampere can be detected.

An automatic balance has a quoted sensitivity of 1 vernier division/ 0.1 mg. This means that the index moves through 1 vernier division when the mass changes by 0.1 mg.

When the input to a measuring system is of the same form as the output the term *magnification* is sometimes used instead of sensitivity.

$$\text{magnification} = \frac{\text{output}}{\text{input}}$$

Magnification will often be found to be the term used in the case of optical instruments, e.g. a microscope, or mechanical devices, e.g. a mechanical comparator such as a dial gauge.

Another term that is used is *gain*. This term is often used with electronic equipment but is not restricted to just electronic systems.

$$\text{gain} = \frac{\text{output}}{\text{input}}$$

Instead of the word 'gain', *amplification* may sometimes be used, but it describes the same thing.

The terms sensitivity, magnification, gain and amplification all describe the relationship between the output and the input and which one is used depends on the type of system being described. If a graph is plotted of the output against the input and the result is a straight line passing through the origin then the output is always directly proportional to the input and the system is said to have a *linear response*.

Where a measuring system can be used to measure slowly or quickly changing quantities it is often necessary to distinguish between the accuracy for the slow change and the accuracy for the fast change. These are often referred to as the *static and dynamic accuracy*. A mercury-in-glass thermometer may have a maximum error of 0.2°C for slowly changing temperatures. However if it were in a bath of liquid with the temperature changing very rapidly the thermometer might well lag sufficiently behind the actual temperature to give an error considerably greater than the 0.2°C static error. Mercury-in-glass thermometers do not respond quickly to changing temperatures, their speed of response is low. This can be put as—the *response time* of the mercury-in-glass thermometer is high. The thermometer takes a long time to respond to a change in temperature.

Ideally a measuring system will give the same result irrespective of whether the quantity being measured is increasing from a lesser value or decreasing from a larger value. Thus if a thermometer were the measuring system it would ideally give the same temperature reading irrespective of whether it was cooling or warming to the temperature concerned. A normal laboratory mercury-in-glass thermometer being used to measure a temperature of, say, 60°C will not give the same reading, at least for some quite significant time, if the thermometer has been initially at 20°C and introduced into the liquid at 60°C or if it has been initially at 100°C and then introduced into the liquid. Such a measuring system is said to exhibit *hysteresis*.

CALIBRATION

Calibration is the process of checking a measuring system against a standard or putting marks on a display when the transducer is in a defined environment. This may take the form of comparison with:

(1) A primary standard.
(2) A secondary standard having an accuracy greater than the system being calibrated.
(3) A known input source which involves other measurements which can be made to a greater accuracy than the system being calibrated.

The primary standard of mass is an alloy cylinder (90% platinum—10% iridium) of equal height and diameter, held at the International Bureau of Weights and Measures at Sèvres in France. The mass is defined as one kilogram. Duplicates of this standard are held by other countries. In Great Britain the duplicate mass standard is kept by the National Physical Laboratory.

The primary standard of length is the metre and is defined as a length equal to 1 650 763.73 wavelengths in vacuum of a particular radiation band emitted by the krypton-86 atom.

The primary standard of current is the ampere and this is defined as duration of 9 192 631 770 periods of oscillation of the radiation emitted by the caesium-133 atom under precisely defined conditions of resonance.

The primary standard of current is the ampere and this is defined as that constant current which, if maintained in two straight parallel conductors of infinite length, of negligible circular cross-section, and placed one metre apart in a vacuum, would produce between these conductors a force equal to 2×10^{-7} N per metre of length.

The primary standard of temperature is the kelvin (K) and this is defined so that the temperature at which liquid water, water vapour and ice are in equilibrium (known as the triple point) is 273.16 K.

The primary standard of luminous intensity is the candela and this is defined as the luminous intensity, in the perpendicular direction, of a surface of $1/600\,000$ m^2 of a black body at the temperature of freezing platinum under a pressure of $101\,325$ N/m^2.

In practice these primary standards are rarely used for calibration but more practical, secondary, standards are used. These all originate from the primary standards. Thus for calibration of a measuring system for length in a workshop or laboratory it is most likely that a secondary standard will be used rather than the primary standard of length. The primary standard is not very convenient for a working standard. An accurately made steel rule may be a suitable secondary standard of length in some instances.

In the case of temperature a secondary standard may be a mercury-in-glass thermometer that has been carefully calibrated by the National Physical Laboratory (NPL) or the British Calibration Service (BCS). A thermometer is calibrated against their standard and a calibration certificate issued. For Great Britain the National Physical Laboratory has the responsibility for primary standards. Both the NPL and the BCS carry out calibrations for industry.

For time measurements a secondary standard is time signals broadcast by radio. Another possibility is a watch calibrated by the National Physical Laboratory and issued with a calibration certificate. A further possible secondary standard is the frequency of the mains alternating current, 50 Hz. Such a signal could be used as a known input source to a measuring system. An adequate laboratory standard may be a quartz crystal watch.

A calibration chart for a particular instrument can take the form of a table in which the readings of the instrument at certain specified points within the instrument range are shown in comparison with the standard values. Another form is a graph relating the instrument readings with the results from the standard.

Many measuring systems use transducers which give a response which is not perfectly proportional to the input signal and, though such a system may be calibrated at, say, two points, there can be some inaccuracy in assuming that it is a perfectly linear scale for readings between the calibration points. Thus for a perfectly linear scale a reading half-way between two calibrated points would be expected to have a value exactly half-way between the two calibration points. If the system is not perfectly linear such an assumption will lead to some inaccuracy. In the case of a mercury-in-glass thermometer calibrated at the 0°C and 100°C points a temperature of 50°C will generally not be exactly half-way between the two calibration points because the expansion is not perfectly linear.

There are a number of reasons why a measuring system in use may not conform to its calibration. One obvious reason is that the system is being used under different conditions to those prevailing when the instrument was calibrated. Most measuring systems are temperature sensitive and calibration is generally for just one specified temperature. The most common temperature used for calibration is 20°C, this being a temperature relatively close to the normal conditions under which most measuring systems are used. Other environmental conditions can also affect a system, e.g. some are affected by changes in atmospheric pressure, some by humidity. Another reason for a measuring system not conforming to its calibration is if the system is not correctly set up, i.e. not set up in the same way as that used for its calibration. Most systems

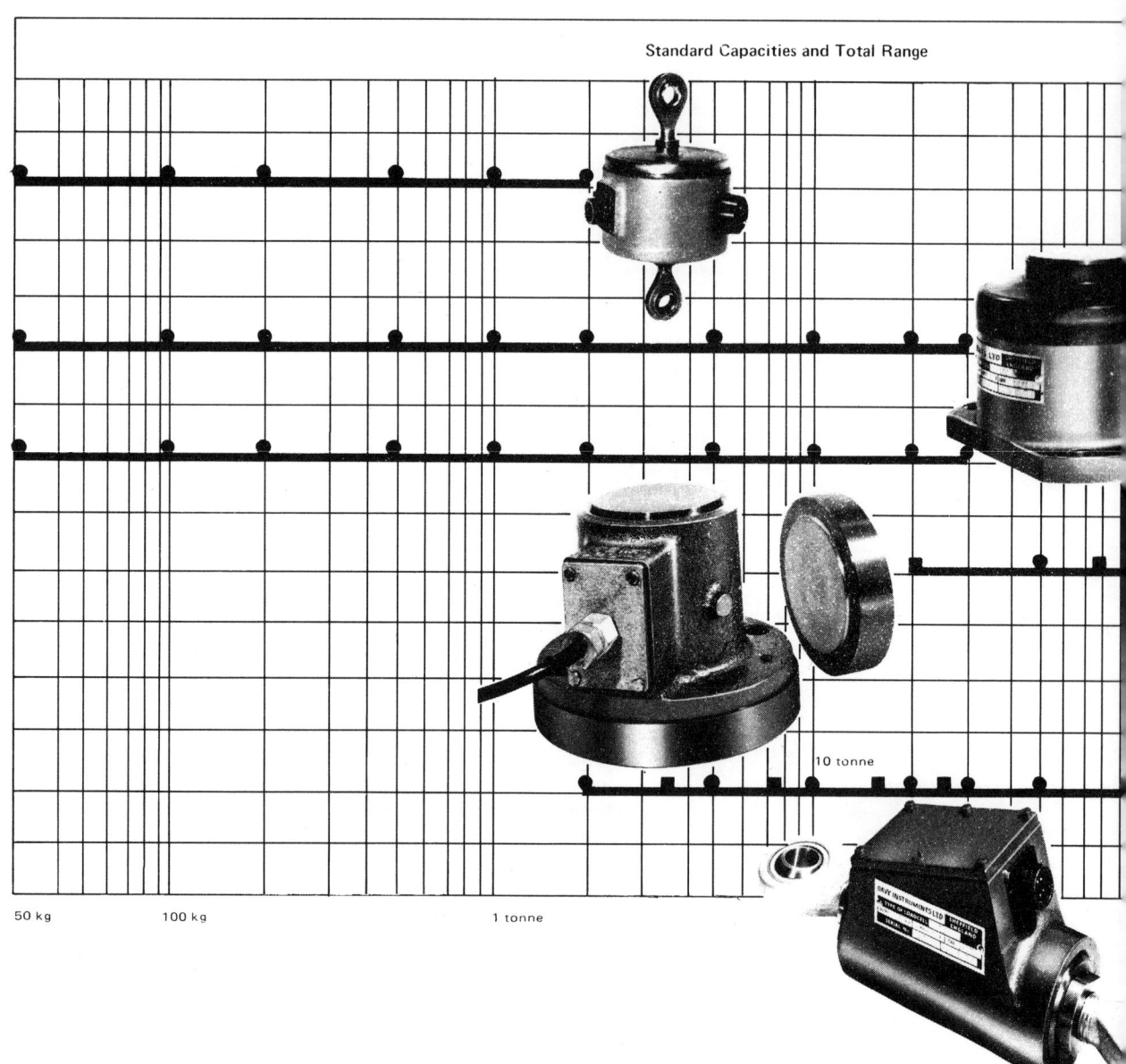

Standard Capacities and Total Range

10 tonne

50 kg   100 kg   1 tonne

Figure 1.3 Range of load cells (reproduced by permission of Davy Instruments Ltd)

	Series	Accuracy	Application	Data Sheet
		Total error due to non-linearity, hysteresis and non-repeatability		
Tension	TMA	± 0.1%	Low range, high accuracy weighing and force measurement	LD 210
Compression	EM/EMB	± 0.25%	Low-medium accuracy container weighing; bin-level indication	LD 208/3
Compression	EMP/EMA	± 0.1%	High accuracy weighing	LD 208/3
Compression	HM	± 0.1%	High accuracy 'heavy' weighing and force measurement	LD 203/3
Compression	RM	± 0.2/0.5%	Heavy-machinery force measurement	ED 103/3
Tension	TM	± 0.05%	Highest accuracy weighing and force measurement. (Also available as tension-compression units).	LD 206/3

00 tonne   1000 tonne   6000 tonne ▷

All Davy cells are complete with self-aligning loading pads (or self-aligning spade ends on tension cells) which enable them to cope with the effects of slight non-axial loading.

PREFERRED CAPACITIES AVAILABLE AT SHORT DELIVERY

AVAILABLE AS SPECIAL MANUFACTURING ONLY

have zero adjustments and these need to be correctly made for the calibration to be valid.

A particular mercury-in-glass thermometer may be calibrated for use by total immersion in the liquid for which the temperature is required. If such a thermometer is used only partially immersed the calibration will not be valid and there may be a greater error than indicated by the calibration.

A vernier microscope is specified as being calibrated at 20°C. At that temperature the maximum error is quoted as being ± 0.01 mm. If the instrument is used at other temperatures the error can be larger than ± 0.01 mm. The coefficient of expansion of the scale of the instrument is quoted as being 0.000 017 per °C.

$$\text{coefficient of expansion} = \frac{\text{change in length}}{\text{length} \times \text{change in temperature}}$$

Thus if the temperature at which the instrument is used is 30°C, i.e. 10°C above the calibration temperature, then for a scale length of 160 mm

$$\text{change in length} = 0.000\ 017 \times 160 \times 10$$

$$= 0.0272 \text{ mm}$$

The length of the scale changes by more than the maximum quoted error.

Many measuring systems have facilities built in by which their calibration can be checked against some secondary standard.

PROBLEMS

(1)   Define the following terms: transducer, signal conditioner, display element.

(2)   Identify the different elements, i.e. transducer, signal conditioner and display element, in an extensometer measuring system.

(3)   Identify the different elements in a time measuring system such as a pendulum clock.

(4)   Define the following terms as applied to a measuring system: range, accuracy, reproducibility, repeatability, constancy, response time, sensitivity.

(5)   A vapour pressure thermometer is advertised as having ranges of −15 to 35°C and 20 to 120°C with a maximum error of 1% of the full scale reading. If the thermometer were used to measure a temperature of (a) 10°C and (b) 60°C, within what spread of temperature could the instrument be expected to give readings?

(6)   A thermometer having a maximum error of 0.4°C gives a reading of 45.8°C. What can be said about the actual temperature?

(7)   A hydrometer has a range of 600 to 700 kg/m³ and a maximum error of ± 2 kg/m³. If the instrument indicates a density of 650 kg/m³, within what spread of density would the actual density be expected to lie? Note: Hydrometers are used for the measurement of the densities of liquids.

(8)   The sensitivities of two thermocouples are quoted as follows:

copper/constantan thermocouple   0.03 millivolts/°C

iron/constantan thermocouple   0.05 millivolts/°C

Which thermocouple will give the larger voltage for a given change in temperature?

(9)   The amplifier system of a cathode ray oscilloscope is said to have a sensitivity of 0.1 V/cm. What would be the deflection on the screen for an input of 0.3 V?

(10)   A spot galvanometer is quoted as having a sensitivity of 15 mm/μA. The scale is 160 mm long and marked in millimetres. What is the largest current that can be indicated by the instrument?

(11)   A measuring system is said to have a gain of 10. What does this mean?

(12)   A vernier microscope is said to have a magnification of 30. What does this mean? (Note: With microscopes the magnification would usually be written as $\times 30$.)

(13)   *Figure 1.3* is reproduced from a manufacturer's data sheet for load cells. A load cell is a measuring system that can be used for the measurement of forces or loads (see Chapter 5).

(a) Which load cells could be used for a measurement of load of about 100 tonne?

(b) The accuracy of the load cells is given in terms of 'total error due to non-linearity, hysteresis and non-repeatability'. Explain these terms and why they contribute to the total error.

(c) The accuracy for the TMA series load cell is quoted as $\pm 0.1\%$. How close to the true value could such a cell be expected to give a result if the load being measured was about 100 kg?

(d) Which load cell would be suitable for the measurement of a load of about 10 tonnes to an accuracy of ± 0.005 tonne?

(e) Which load cell would seem the most appropriate for the very rough measurement of a load of about 1 tonne if the result only needs to be known to within about 0.005 tonne?

(f) Which data sheets would you request from the manufacturer if you were considering ordering a load cell to perform the measurements indicated in (d) and (e)?

2 Transducers

Aims: At the end of the chapter you should be able to:
Explain the function of transducers.
Describe the principles of operation of common transducers used in the measurement of temperature, displacement, strain and the detection of discrete events.

(Chapter 5 looks in more detail at measuring systems using the above transducers.)

WHAT ARE TRANSDUCERS?

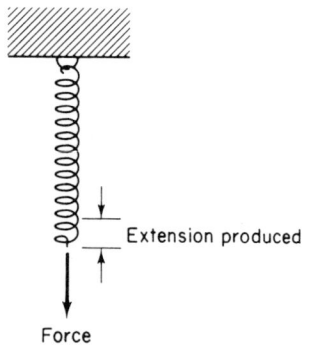

Figure 2.1 A spring is a transducer

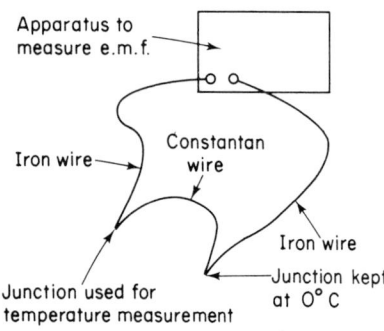

Figure 2.2 A thermocouple is a transducer

The term *transducer* is used to describe any item which changes information from one form to another. The reason for the change is to obtain information which can easily be measured.

A simple example of a transducer is a *spring*. If a force is applied to a tethered spring (*Figure 2.1*) it stretches. Information about the force is changed to information in the form of a displacement of the end of the spring and using this principle a measurement of the displacement is used as a measure of the force in the simple spring balance. Different forces give rise to different displacements and so a measurement of the displacement gives a unique indication of the force. For a simple spring the displacement is proportional to the force

displacement $x \propto$ force F

This is often written as an equation

$F = kx$

where k is a constant.

A *thermocouple* (*Figure 2.2*) is a transducer which converts information about a temperature difference to information in the form of an e.m.f. Measurement of the e.m.f. can be used as a measure of temperature. In this case the two variables are not generally directly proportional and either a calibration graph or a reference table is used. The following is part of a reference table for an iron-constantan thermocouple.

e.m.f. in microvolts when the cold junction is at $0°C$, temperature in $°C$.

temperature	0	10	20	30	40	50	60	etc.
e.m.f.	0	500	1020	1540	2060	2580	3110	

A *potentiometer* is a transducer in which a rotation or displacement is converted into a potential difference (*Figure 2.3*). The potentiometer consists of a resistance across which a potential difference is maintained. The output potential difference is obtained between one end of the resistance and slider which can move, in electrical contact, along the resistance element. The position of the slider along the resistance element determines the size of the output potential difference. If the movement of the slider is in a straight line along a linear resistance then displacement information is converted into information in the form of a potential difference. The output can be arranged to be directly proportional to the displacement. If the movement of the slider is in a

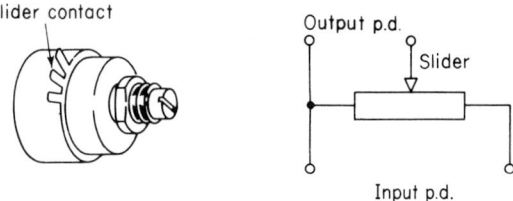

Figure 2.3 A potentiometer is a transducer

circular path along a resistance element formed on the circumference of the circle then rotation information is converted into information in the form of a potential difference. The output can be arranged to be directly proportional to the angular rotation.

  These are just three examples of transducers. The remainder of this chapter considers a few of the more common transducers in more detail. All however have the same common feature: they convert information from one form to another.

TRANSDUCERS FOR TEMPERATURE MEASUREMENT

A change in temperature of a body can result in a variety of other changes. For example there could be:

(1) A change in dimensions, i.e. expansion or contraction, of material in the form of solid, liquid or gas.
(2) A change in electrical resistance, generally for metals and semiconductors.
(3) A thermoelectric e.m.f. for two different metals joined together.
(4) A change in the intensity and colour of the radiation emitted by the hot body.

  There are many forms of thermometers based on the above changes. *Figure 2.4* shows one form of a thermometer based on the expansion of solids, where the thermometer is basically just a *bimetallic strip*. The strip consists of two metals welded or riveted together so that they must both move in unison. The two metals used have different coeffic-

(a)   (b)

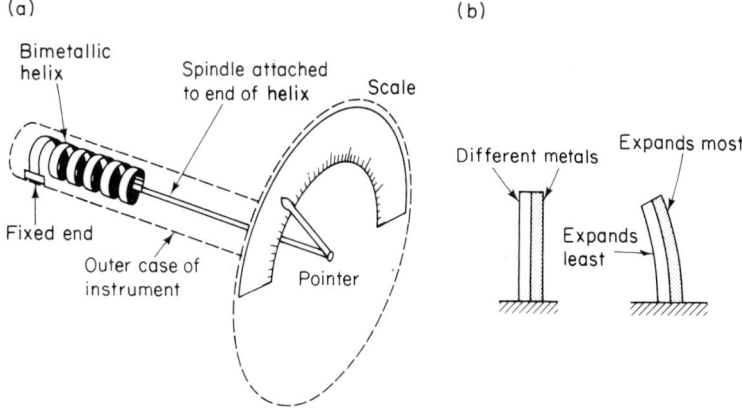

Figure 2.4 (a) A bimetallic thermometer (b) A bimetallic strip

ients of thermal expansion, i.e. the same temperature rise results in the two metals expanding by different amounts. The only way this can happen for the combined strip is for the strip to bend. In the thermometer the strip is wound in the form of a helix and so changes in temperature result in rotation of the free end of the helix. Thus information about temperature is changed to information in the form of expansion and so to rotation and movement of a pointer across a scale. The following is taken from a manufacturer's catalogue and indicates the types of specification available for such bimetallic thermometers.

Ranges for which instruments are available

$-30°$ to $60°C$; $0°$ to $160°C$; $10°$ to $120°C$; $50°$ to $250°C$.

Maximum error of less than 1% of the total scale range.

Graduations are almost linear throughout the chosen range.

The thermometers are claimed to have certain advantages over liquid-filled thermometers in being easier to read and more robust.
    If such a thermometer were used for a specified range of $10°$ to $120°C$ then the maximum error would be at least 1% of the range of $120 - 10 = 110°$, i.e. $1.1°C$. Any reading given by the thermometer would be within about $1°$ of the correct result, i.e. $\pm 1°C$.
    The resistance of a metal wire depends on temperature. Thus it is possible to take a metal wire resistor and use a measurement of its resistance as a measurement of temperature. A typical industrial form of such a *resistance thermometer* might use a nickel wire resistance element. Different shapes of resistance elements are available. The resistance element is at the end of a cable and, if intended for pressing against a solid surface to determine the surface temperature, is a thin flat coil to ensure maximum contact with the surface. The resistance element for use with a liquid is in the form of a rod. The resistance element is chosen to give maximum contact between the resistance element and the object for which the temperature is required.
    The following is a typical specification of a portable resistance thermometer:

Temperature ranges $-30°$ to $+30°C$; $+20°$ to $+80°C$; $+60°$ to $+180°C$.

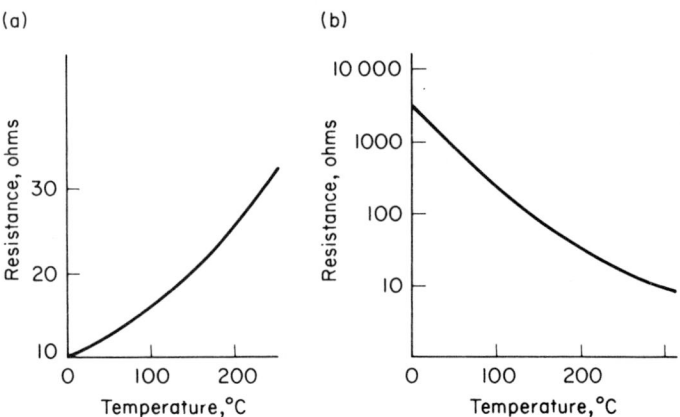

Figure 2.5   (a) Nickel (b) Thermistor; note the log scale

The probe has a nickel resistance element adjusted at 20°C to give readings accurate to ±0.25°C.

Metal resistance elements have the disadvantage that the changes in resistance are comparatively small. The nickel resistance element referred to above has a temperature coefficient of resistance of about 6×10^{-3} °C^{-1}. This means that the change in resistance for an element having an initial resistance of 1 Ω is 6×10^{-3} Ω for a 1°C change in temperature. Semiconductors give much larger changes in resistance for the same temperature change. Semiconductor resistors used for temperature measurement are called *thermistors. Figure 2.5* shows how the resistance of a typical thermistor and typical metal resistor element change with increasing temperature. The resistance of the metal increases with temperature, the increase per degree being quite small. The resistance of the thermistor decreases with increasing temperature, the decrease per degree being much larger than the corresponding change for the metal. Another difference is that the resistance of the metal changes reasonably uniformly with temperature and so gives a linear scale, while the resistance of the thermistor does not change uniformly with temperature and so gives a non-linear scale. One of the problems with thermistors, not present with metal elements, is that an instrument calibrated for use with one thermistor will often need recalibration if used with a replacement thermistor. Thermistors have an advantage over metal resistors in that they have a higher resistivity and so can be made very small and still have a high resistance. Their size enables them to be used more easily for temperature measurements over a small area. The smaller mass results in the element more quickly attaining temperatures and they are thus useful for measurements where temperatures are rapidly changing. *Figure 2.6* shows some typical forms of thermistor.

The *thermocouple* consists of two dissimilar electrical conductors joined together. When the two materials are part of the circuit with a measuring instrument (see *Figure 2.2*) there are two junctions and if there is a temperature difference between the two junctions then an e.m.f. is set up. The size of the e.m.f. depends on the difference in temperature and the materials involved.

Copper and constantan are useful metals for a general purpose thermocouple in the temperature region of about −100°C to +400°C.

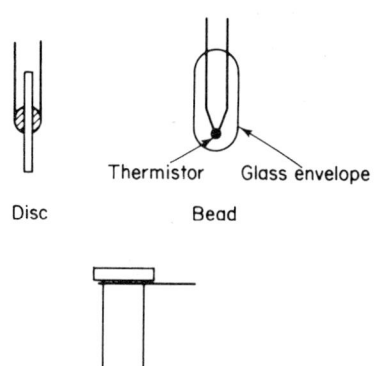

Disc

Thermistor   Glass envelope

Bead

Rod

Figure 2.6   Some forms of thermistor

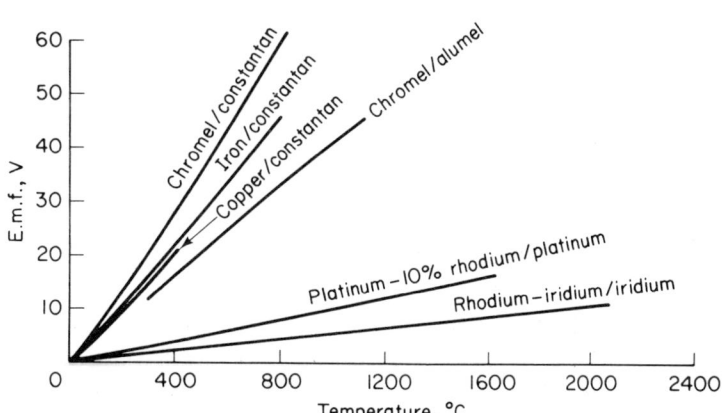

Figure 2.7

Iron and constantan can be used from about $-100°C$ to about $800°C$. A nickel-chromium alloy (chromel) and a nickel-aluminium alloy (alumel) can be used from about $-200°C$ to $+1200°C$. Platinum with a platinum-rhodium alloy can be used from about $0°C$ to $1600°C$. Other combinations of metals can be used to even higher temperatures. *Figure 2.7* shows how the e.m.f. of some thermocouple materials varies with temperature, one of the thermocouple junctions being at $0°C$.

Thermocouples enable the temperature to be measured over quite a small area, the area of the junction between the two metals. Because they have a very small mass they can respond very rapidly to temperature changes.

Figure 2.8 shows an example of an instrument that depends on the radiation emitted by a hot object, called a *disappearing filament pyrometer*. The intensity of the light emitted by the hot object is compared

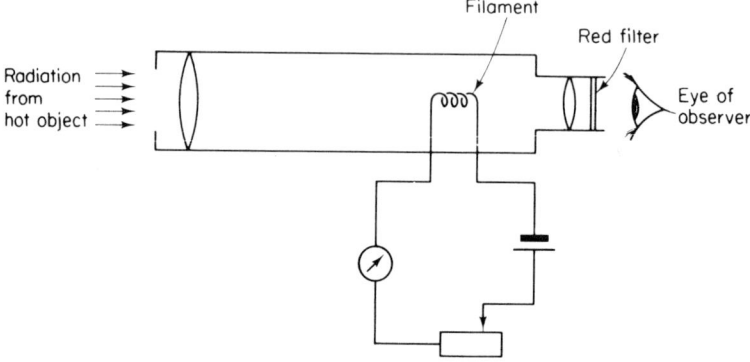

Figure 2.8   The principle of the disappearing filament pyrometer

with the intensity of the light emitted by a tungsten filament. The intensities are generally compared for just a small part of the spectrum, the hot objects being viewed through a red filter. The filament is heated electrically by a current from a battery. The current is controlled by a variable resistor which is adjusted until the intensity of the light emitted by the hot filament is the same as that from the hot object. When this occurs the filament merges into the background of the hot object and seems to disappear. The current for which this occurs is thus a measure of the temperature and can therefore be calibrated in terms of temperature.

In the disappearing filament pyrometer information about the temperature of the hot object is matched with information about the temperature of the hot filament which is then converted into information in the form of a current.

TRANSDUCERS FOR DISPLACEMENT MEASUREMENTS

The simple *potentiometer* described earlier (*Figure 2.3*) is a displacement transducer in which displacement information is changed into information in the form of a potential difference.

Displacement information → p.d. information

Figure 2.9 shows another simple displacement transducer, called a *pneumatic comparator*. Displacement of the block, i.e. changes in the

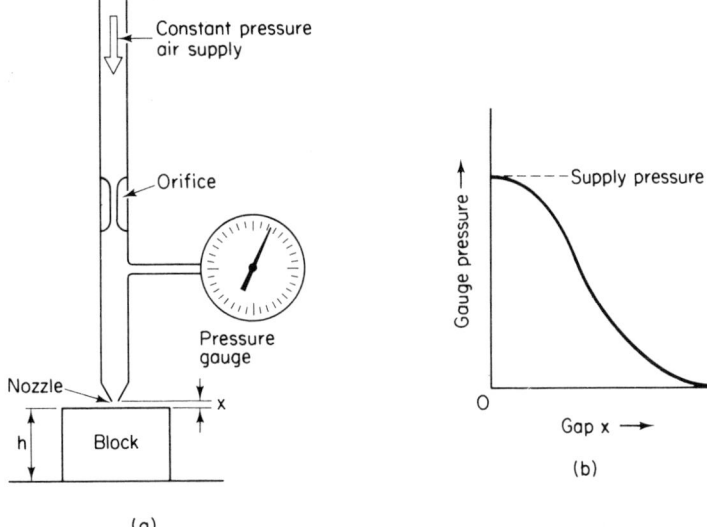

Figure 2.9
(a) The basic features of a pneumatic comparator
(b) The typical form of a calibration graph

height h, result in changes in the pressure reading given by the pressure gauge.

Displacement information $\rightarrow$ pressure information

Air from a constant pressure supply flows through the orifice and out of the nozzle. The pressure gauge measures the pressure between the orifice and the nozzle. If there is no blockage of the flow of air out of the nozzle then the pressure between the orifice and the nozzle, i.e. x = infinity is close to the atmospheric pressure. The relative sizes of the orifice and nozzle are chosen so that this occurs. If the nozzle is completely blocked, i.e. x = 0, then the pressure indicated by the gauge rises to the same as the pressure of the air supply before it passes through the orifice. The pressure indicated by the gauge thus depends on the value of x.

Another form of displacement transducer is the *inductance transducer*. When an alternating current flows in a coil, alternating e.m.f. can be induced in a neighbouring coil. The effect is called electromagnetic induction. The size of the induced alternating e.m.f. depends on the closeness of the two coils and the amount of magnetic material present. Thus if the two coils are wound side-by-side on an iron core then the induced e.m.f. is much larger than if there were no iron core present and the two coils were just side-by-side in air.

Figure 2.10 shows two forms of inductance transducer based on this principle. In *Figure 2.10a* the movement of the plunger causes an iron core to move along the axis of two coils, which are mounted one on top of the other. The two coils are connected in a bridge circuit so that when the iron core is equally in each coil there is no voltage indicated on the voltmeter. When the iron core is more in one coil than the other the voltmeter indicates a voltage. Thus information about the position of the iron core is changed to information in the form of a voltage.

Displacement information   →   p.d. information

Figure 2.10b shows another form of inductance transducer, known as the *linear variable differential transformer* (LVDT). This has three coils, the two outer coils being connected together. The centre coil is supplied with an alternating current. This induces e.m.f.s in the two outer coils. These coils are connected in such a way that when the same e.m.f. is induced in each coil the e.m.f.s cancel each other out and there is no resultant potential difference across the two coils. Only when the e.m.f. induced in each of the outer coils is different is there an output potential difference. The e.m.f. induced in each coil depends on the position of the iron core within each coil, thus movement of the core so that one coil contains more iron core than the other results in an

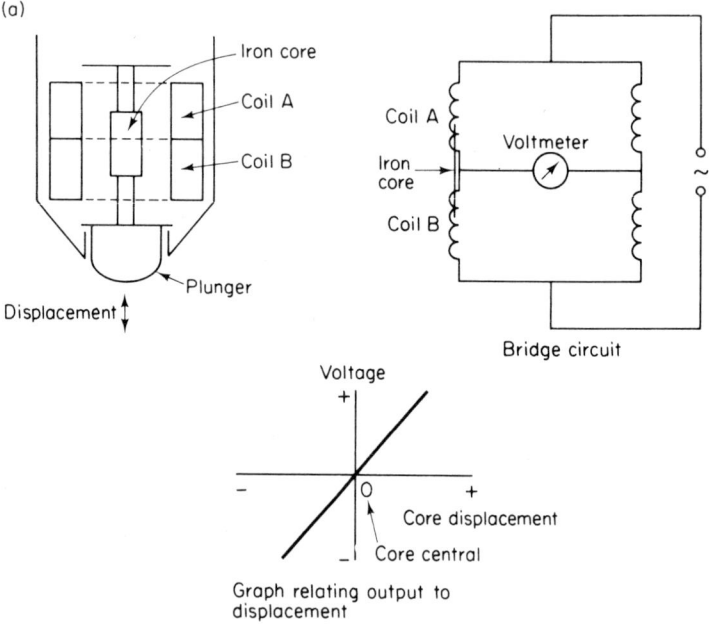

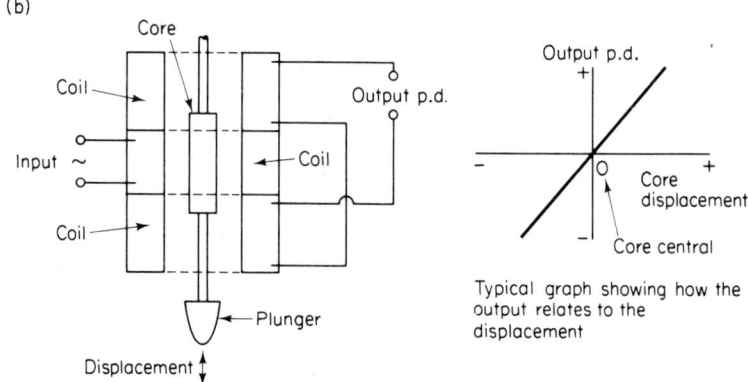

Figure 2.10
(a) Bridge type inductance transducer
(b) Linear variable differential transformer

output potential difference. The iron core can be linked to the item whose displacement is required and so displacement information is converted into information in the form of a potential difference. For quite a reasonable range of core displacement the output potential difference is directly proportional to the displacement, hence the word 'linear' in the name given to this transducer.

TRANSDUCERS FOR STRAIN MEASUREMENTS

If a piece of metal wire is stretched not only does it get longer and narrower, but its resistance increases. The greater the strain experienced by the wire the greater the increase in resistance. The electrical resistance strain gauge is based on this effect. Information in the form of a change in length is converted to information in the form of a resistance change. The *strain gauge* is a transducer.

Figure 2.11 shows two forms of strain gauge. One is made from metal foil, the other from metal wire. Both resistance elements are mounted on an insulator backing so that when they are attached to the surface of a metal they are electrically insulated from the metal.

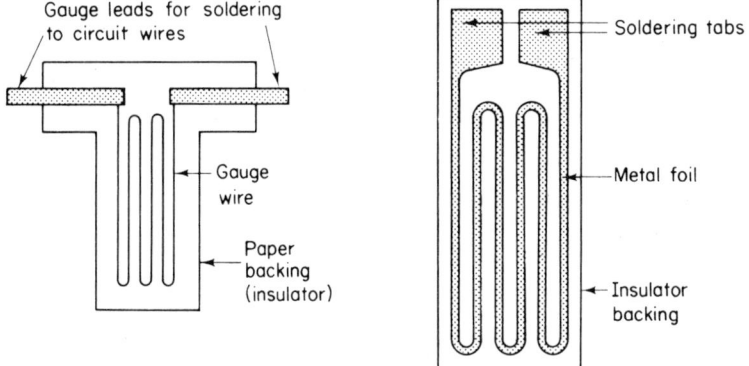

Figure 2.11   Strain gauges

In use the strain gauges are bonded to the surface of the component for which the surface strain is required. It is important that the bonding is done carefully as when the surface is strained the strain gauge must also be strained and not slip. To ensure good bonding the surface must be carefully prepared—generally slightly roughened and then degreased. Then the adhesive prescribed by the manufacturer must be used according to the instructions and the adhesive drying time elapse before the gauge is used.

Measurements of the resistance change occurring when the mounted gauge is strained can be used to enable the strain to be calculated. The relationship between the strain and the change in resistance ΔR is given by

$$\frac{\Delta R}{R} = G \times \text{strain}$$

where R is the gauge resistance and G a constant called the gauge factor. For most metal strain gauges the gauge factor has a value of about 2. The gauge factor is supplied by the strain gauge manufacturer from a calibration made of a number of gauges taken from the same production batch.

A strain gauge with a resistance of 100.0 Ω and having a gauge factor of 2.0 is subject to strain which produces a change in resistance of 0.004 Ω. What is the strain acting along the axis of the gauge? Using the above equation gives

$$\frac{0.004}{100.0} = 2.0 \times \text{strain}$$

Hence strain = 2×10^{-5}

As will be seen from the above example the measurement of strain by means of a strain gauge involves the measurement of very small changes in resistance. Some commercial instruments for use with strain gauges are precalibrated to give the readings directly as strain instead of change in resistance.

When the temperature of a resistor changes then its resistance changes. Thus a strain gauge is sensitive to both strain and temperature. The resistance changes produced by temperature changes can be comparable with the resistance changes produced by strain. Thus in the use of a strain gauge for strain measurement the effects of any temperature changes have to be eliminated. This is often done by using what is called a dummy gauge. This is a strain gauge of the same resistance as the gauge being strained, known as the active gauge, and mounted on a piece of the same material as the active gauge. This piece of material is not however subject to the strain. The active and the dummy gauges are connected to a measuring circuit in such a way that the effects of temperature on the two gauges cancel out leaving only the difference between the two resulting from the effect of the strain to be measured.

The strain gauges so far considered in this section use the extension of metal wire or foil strips and have gauge factors of about 2. Strain gauges can be made from semiconductor material. Such strain gauges (*Figure 2.12*) have considerably higher gauge factors, in the range 100 to 200. They also have the advantage that the material has a higher resistivity than metals and so can more easily have a high initial resistance. Both this higher resistivity and higher gauge factor mean that the change in resistance for a given strain can be much higher with a semiconductor strain gauge than with a metal wire or foil strain gauge. Semiconductor strain gauges do however have a disadvantage—they are more temperature sensitive. It is thus essential to compensate for temperature fluctuations when the gauges are used for the measurement of strain.

A transducer which is particularly useful where the strain is changing quite rapidly is based on the *piezo-electric effect*. Certain crystals, such as quartz, Rochelle salt (potassium sodium tartrate) and barium titanate, have the property that when they are subject to a strain along a certain axis complementary electrostatic charges appear on opposite faces of the crystal, i.e. a potential difference has been produced. The crystal record player pick-up depends on this effect. The wavy form of the groove in the record causes one end of a crystal strip to be flexed, the other end being fixed. The crystal suffers a strain which is continually changing. The resulting potential difference thus varies in a manner dictated by the wavy form of the groove and after amplification, is fed to a loudspeaker and produces a sound. The form of the sound is dictated by the form of the potential difference which in turn is dictated by the form of the groove.

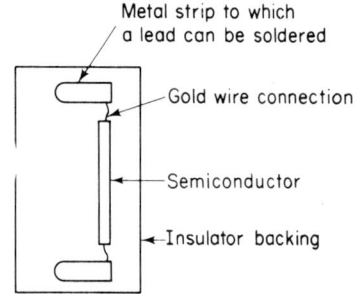

Metal strip to which a lead can be soldered

Gold wire connection

Semiconductor

Insulator backing

Figure 2.12 Semiconductor strain gauge

The size of the potential difference depends on the strain, i.e. the greater the strain the greater the potential difference. If this potential difference is used to produce a current in an external circuit then the potential difference drops, even though the strain remains constant. Only when the strain changes will the potential difference be able to be restored. One of the problems in using piezo-electric transducers is that of measuring the potential difference without drawing current. Because of this, such transducers are more often used for fast changing strains rather than static strains.

TRANSDUCERS FOR DETECTING DISCRETE EVENTS

A toothed wheel rotates—how can the number of teeth passing a point per second be measured? Items move on a conveyor belt—how can the number of items passing a point per second be measured? These are examples of where a transducer is required to detect discrete events happening and enable a count to be made. Such systems are often called *digital systems.*

Figure 2.13 shows a simple *digital transducer.* It is a coil of wire on a bar magnet. If any ferrous material moves close to the magnet a current pulse is produced, the result of electromagnetic induction, which can be used to activate a counter. Another variation is to just have the coil connected to the counter and have the magnet mounted on the object being counted. As the magnet moves past the coil a current pulse is produced, the result of electromagnetic induction, which activates the counter.

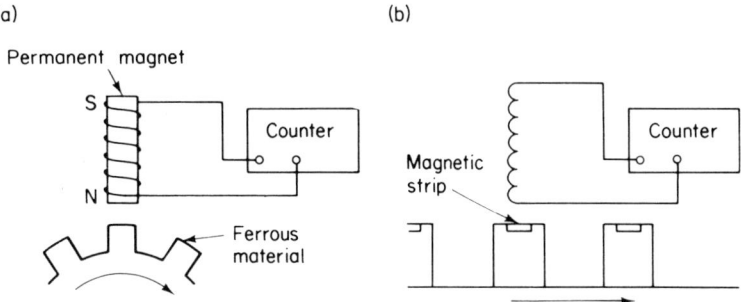

Figure 2.13 Simple inductive transducer

Figure 2.14 shows another form of digital transducer. This is a *photo-electric transducer.* When the light falling on the transducer is interrupted a current pulse is produced which can activate a counter. There are three basic types of photo-electric transducer—photo-emissive, photo-conductive and photo-voltaic. In the *photo-emissive transducer* the light falling on a light sensitive surface causes it to emit electrons which are then attracted to a positively charged electrode and cause a current to flow. Interrupting the light falling on the light sensitive surface causes a change in the current. In the *photo-conductive transducer* the light falling on to the light sensitive substance causes its resistance to change. The change in resistance causes the current in the circuit to change. In the *photo-voltaic transducer* the light falls on two different substances in contact and

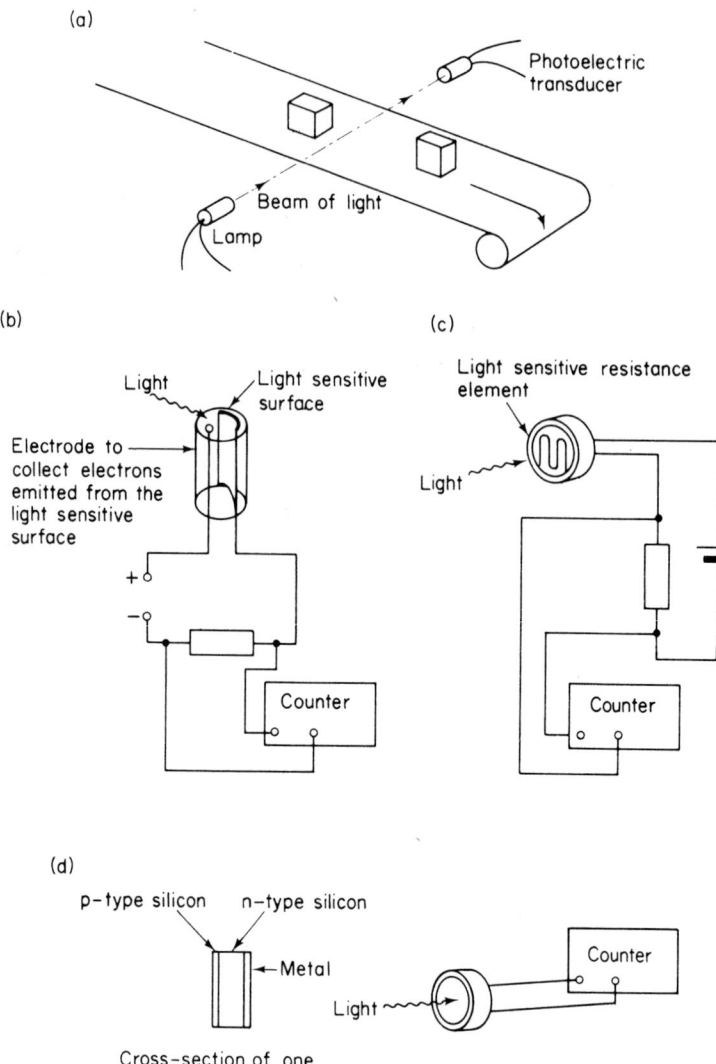

Figure 2.14 Photo-electric transducers
(a) Counting objects using a photo-electric transducer
(b) Photo-emissive transducer
(c) Photo-conductive transducer
(d) Photo-voltaic transducer

causes an e.m.f. to be produced. Interrupting the light causes the e.m.f. to change and hence a change in current in the associated circuit.

Photo-emissive transducer
    light information → current information

Photo-conductive transducer
    light information → resistance information

Photo-voltaic transducer
    light information → e.m.f. information

Photo-electric transducers are not restricted to just digital purposes but can be used wherever measuring systems are required which involve information being received in the form of light. Such transducers offer a great advantage over many other possible transducers in that they do not involve any contact being made with the object being counted, just the interruption of a beam of light. The transducers can be selected to operate with infra-red radiation. The 'light' does not have to be visible.

PROBLEMS

(1) Specify the form of the input and output information for each of the following.
(a) A car speedometer.
(b) A record player pick-up.
(c) A thermocouple.
(d) A hydrometer.
(2) Describe the principle of operation of one of the following.
(a) A thermistor used for temperature measurement.
(b) A bimetallic strip thermometer.
(c) A thermocouple.
(3) Describe the principle of operation of a piezo-electric transducer and explain why it is generally used for dynamic rather than static measurements.
(4) Describe the principle of operation of an electrical resistance strain gauge.
(5) An electrical resistance strain gauge is to be used for the measurement of a strain of 0.1%. If the gauge has a gauge factor of 2.1 and a resistance of 180 Ω what will be the change in resistance of the gauge?
(6) Describe the principle of operation of a linear variable differential transformer.
(7) Describe the principle of operation of one of the following.
(a) Photo-emissive cell.
(b) Photo-conductive cell.
(c) Photo-voltaic cell.
(8) Specify a transducer for each of the following information changes.
(a) Displacement information to potential difference information.
(b) Strain information to resistance information.
(c) Temperature information to resistance information.
(9) Explain the criteria that you would need to adopt to select the pairs of metals to use for a thermocouple. Use *Figure 2.7* to illustrate your answer.
(10) *Figure 2.15* is taken from a manufacturer's brochure and describes a measurement system called a 'bearing wear detector'.
(a) What is the input information to and the output information from the transducer?
(b) Explain the principle on which the transducer operates.
(c) What is the input information to the display element and how is it displayed?

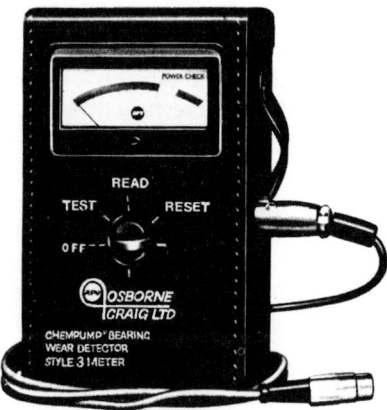

Ex 76004 / B
⟨Ex⟩ ia Ⅱ B. T4

This electronic equipment has been
designed to measure the radial wear on
the rear bearing of a Chempump under
running service conditions, and display
the result on a portable instrument
carried by maintenance personnel.
Within practical limits only one meter is
required for any number of pumps.
Measurement is simple, taking no
more than thirty seconds, the meter
being calibrated to give a prompt
interpretation of the state of the
bearings.
Physically the system comprises two
components:
1. Electro-magnetic transducer
   fitted to pump.
2. Portable measuring
   instrument.
Both items are individually calibrated
to enable any pump to operate with
any measuring instrument.
Direct readings of the bearing
condition is uncomplicated, requiring
little experience on the part of the
operator. Intrinsic safety requirements
were considered during the design of
the electronic circuits, permitting
approved assemblies to be installed in
hazardous environments to SFA 3012
Group 11 B classification.
To detect the eccentric motion of the
shaft and thus the bearing wear, a
transducer and shaft extension are
added to the standard pump. The
standard rotor assembly is replaced by
one of the same diameter but extended
in length to project outside the normal

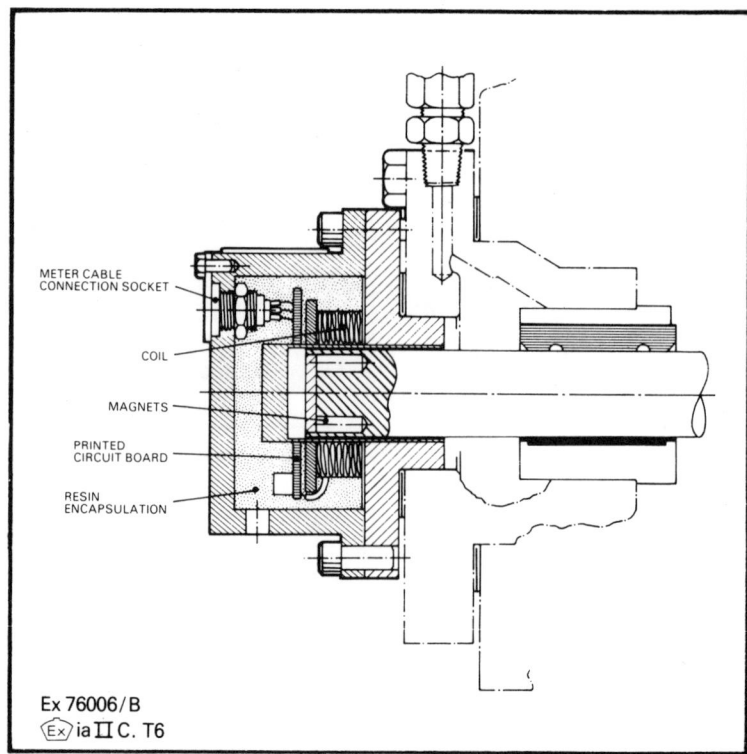

Ex 76006 / B
⟨Ex⟩ ia Ⅱ C. T6

profile at the rear of the pump. The
existing pump rear bearing housing is
replaced by a housing with a thin-wall
stainless steel chamber projecting over
the shaft extension. A coil is carefully
wound around the chamber to produce
concentric turns, each cutting as little
flux as possible when the shaft rotates
concentrically.

Uneven magnetic fields are formed
when the shaft runs 'off centre' due to
bearing wear and causes the turns of
the coil to cut more lines of flux and
hence generate a potential at the
terminals of the coil.
Interchangeability of pump modules is
made possible by the provision of a
calibration control to ensure that each
coil generates the same potential for a
given movement of the shaft
extension. To permit each transducer
to be used in hazardous or non-
hazardous conditions, coils are
protected by resistive and semiconductor components to restrict the
energy at the terminals to a safe level.

A metal enclosure protects the coil and
the electronic components, and
provides support for a connector to the
portable instrument.
A small hand-held battery-powered
instrument provides the check and
measurement functions of the bearing
wear system.

Figure 2.15 Bearing wear detector

Reproduced by permission of
A.P.V.-Osborne Craig Ltd
Glenburn Road
College Milton North
East Kilbride
Glasgow G74 5BJ

World patents
UK   1.480.848
USA   3.981.621
Italy   1022672
Switzerland 576135
Others pending

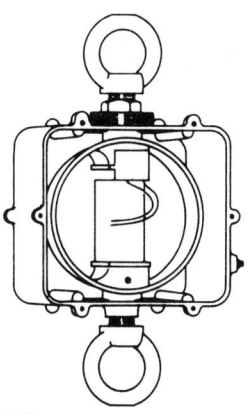

Figure 2.16 Inductive load measuring cell (reproduced by permission of Vibrometer Ltd)

(11) *Figure 2.16* is taken from a manufacturer's catalogue describing a load measuring system based on the use of a linear variable differential transformer. The specification accompanying the picture is as follows:

Inductive load measuring cells

An inductive displacement transducer (LVDT principle) is used to provide a continuous output signal proportional to applied load.

Supply voltage 48–60 V, 50–500 Hz

Rated output   Approx. 1 V

Linearity   ± 1% of rated load

Temperature range −20°C to + 90°C

Temperature influence ± 0.05% per °C on sensitivity
                      ± 0.08% of nom. load per °C on zero

Rated loads in kg 100, 250, 500, 1000, 3000, 5000, 10 000
                   20 000, 30 000, 50 000

(a) Explain how the measuring cell works.
(b) What is the maximum error due to non-linearity when the cell designed for the load range 0–1000 kg is used?

ASSIGNMENTS

(1) Determine the gauge factor of a sample of the strain gauges in a packet. What is the scatter in the results about the value of the gauge factor given by the manufacturer?

(2) Investigate a measuring system used in, say, some industrial process and determine the transducer used. What factors appear to have determined the transducer used?

3 Signal conditioners

Aims: At the end of this chapter you should be able to:
Describe the principle of operation of common signal conditioners.
Identify the signal conditioners within measurement systems.
Consider the ways in which the sensitivity of a measurement system can be changed by changes to the signal conditioner.

WHAT ARE SIGNAL CONDITIONERS?

The element in a measuring system which converts the signal from the transducer into a form which can be displayed is called the *signal conditioner.* Many transducers produce signals which are too small or not in the right form to actuate the display and thus the function of the signal conditioner is to modify the size and form of the signals for the display.

Where an electrical signal is the output from the transducer the signal conditioner may be an *amplifier* which makes the signal large enough to register on a meter. An amplifier is a signal conditioner giving an output larger than the input. Sometimes the amplifier may be built into the display unit, e.g. in the case of a cathode ray oscilloscope. The ratio of output to input for an amplifier is generally called the *gain* of that amplifier (*Figure 3.1a*).

$$\text{Gain} = \frac{\text{output}}{\text{input}}$$

A *lever* is an example of a system which might be used as the signal conditioner where a mechanical displacement needs to be made larger before display. Depending on the relative distances from the pivot of

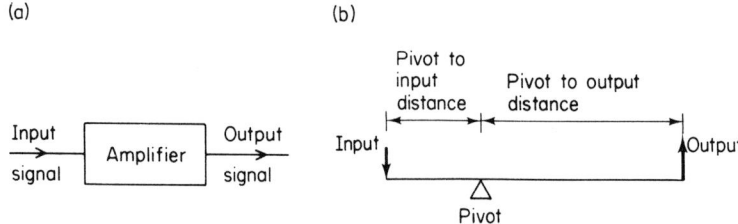

Figure 3.1 Signal conditioners
(a) An amplifier signal conditioner (b) A lever signal conditioner

the application of the displacement signal from the transducer and the output to the display, so the gain, or magnification of a lever can be varied (*Figure 3.1b*). For a single lever

$$\text{Gain} = \frac{\text{output}}{\text{input}} = \frac{\text{distance from pivot that output is taken}}{\text{distance from pivot of input application}}$$

If the distance of the output from the pivot is greater than the distance of the pivot from the input then the gain is greater than one.

When the temperature of a *thermistor* changes, its resistance changes —the thermistor is a transducer. This resistance change has to be made into a form which can be displayed. A simple *electrical circuit* that

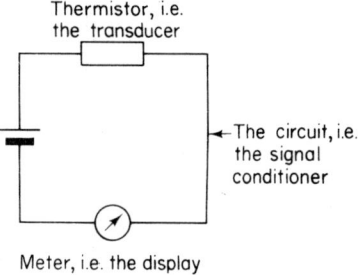

Figure 3.2 A temperature measuring system

would do this is shown in *Figure 3.2*. The changing resistance of the thermistor results in a change in current in the circuit. The changing current is then displayed on a meter. The circuit is the signal conditioner, transforming a resistance change into a current change.

LEVERS   *Figure 3.1b* showed a simple lever. If the pivot to output distance were, say, 100 mm and the distance of the input from the pivot 5 mm then the magnification of the lever would be 100/5 = 20. This means that for every millimetre movement at the input the output end of the lever moves by 20 mm.

Figure 3.3 shows another form of lever. In this case both the input and output points are on the same side of the pivot. The magnification

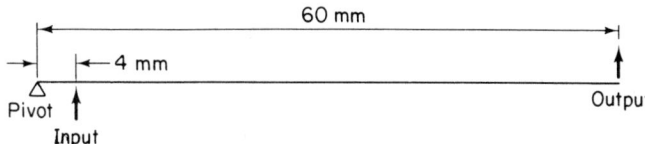

Figure 3.3 A simple lever

is still the ratio of the distance of the output from the pivot to the distance of the input from the pivot. For the distances given in the diagram the gain is 15. Thus a movement at the input of 1 mm results in a movement at the output of 15 mm. A movement at the input of 0.2 mm would mean a movement at the output of 3.0 mm.

Figure 3.4 shows a simple form of *extensometer*. What is the magnification, i.e. how much greater is the movement of the tip of the pointer across the scale than the movement of the point in contact with the specimen being strained? The distance of the tip of the pointer from the pivot is six times the distance of the point in contact with the specimen from the pivot. The magnification is thus 6. Hence an extension of the gauge length by 2 mm would show as a movement of the tip of the pointer across the scale of 12 mm.

To increase the magnification of the extensometer shown in *Figure 3.4* would need either the input distance from the pivot to be decreased

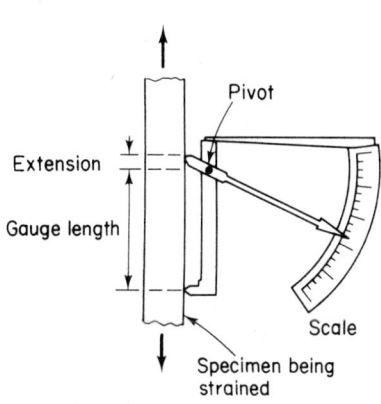

Figure 3.4 A simple extensometer

or the distance of the pointer tip from the pivot to be increased, or both. To obtain a magnification of 10 it would be necessary for the pointer tip to pivot distance to be ten times greater than the input to pivot distance.

Another way of increasing the magnification is to use a *compound lever*. *Figure 3.5* shows a compound lever. The input movement causes one lever to rotate, the movement of the far end of that lever being a magnified version of the input. This magnified movement then causes

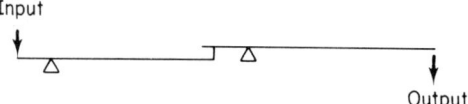

Figure 3.5   A compound lever

another lever to rotate and result in an even more magnified output. If the magnification of the first lever were six and that of the second lever four then the output from the first lever would be six times the input and so the output from the second lever four times six times the input. The overall magnification would thus be 24.

Figure 3.6 shows an *aneroid barometer*. Such an instrument is used to give a response related to the atmospheric pressure. The transducer is the evacuated capsule. When the atmospheric pressure changes the

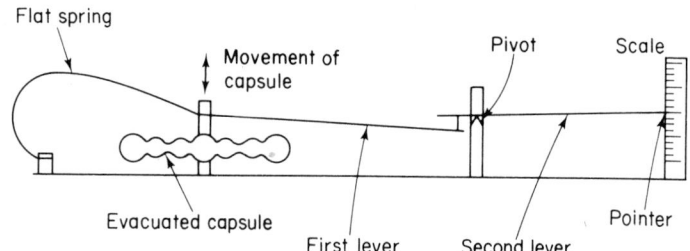

Figure 3.6   An aneroid barometer

capsule expands or contracts under the effect of the changing pressure on its surfaces. This movement is transmitted, and in doing so magnified, to a pointer moving across a scale. The signal conditioner is a compound lever. The first lever pivots about the curved part of the spring and as the input from the transducer is applied closer to this pivot than the output there is a magnification. The end of the first lever transmits this magnified movement to a second lever which again magnifies the movement. The result is a greatly magnified version of the input.

Figure 3.7 shows the *Huggenberger extensometer*. This uses a compound lever to give very high magnifications, often 2000 or more. Extension of the specimen between the gauge points results in rotation of the first lever about its pivot point. The input to this first lever is very close to the pivot point and the output a comparatively large distance away at the far end of the lever. The result is a large magnification of the input movement which is then transmitted to another lever, at a point close to its pivot. The result is a considerably larger movement of the tip of the pointer across the scale.

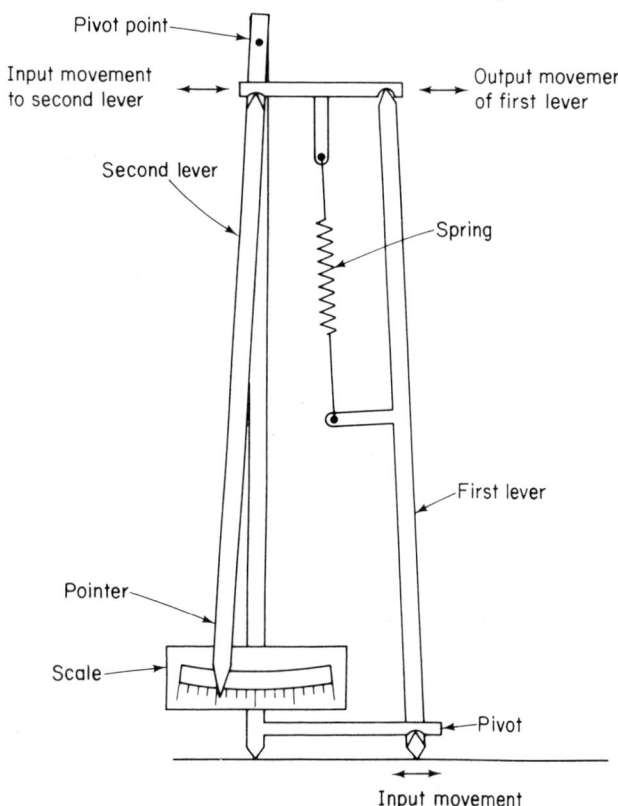

Figure 3.7 Huggenberger extensometer

Figure 3.8 shows the *Sigma comparator.* The signal conditioner is a compound lever. The magnification of the first lever is y/x. The magnification of the second lever is R/r. Thus the overall magnification of the compound lever is

$$\text{magnification} = \frac{y}{x} \times \frac{R}{r}$$

Magnification in the region of 300 to 5000 can be achieved. Movement

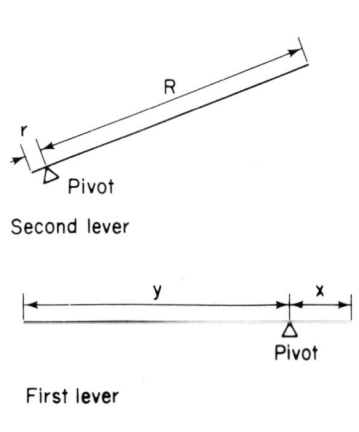

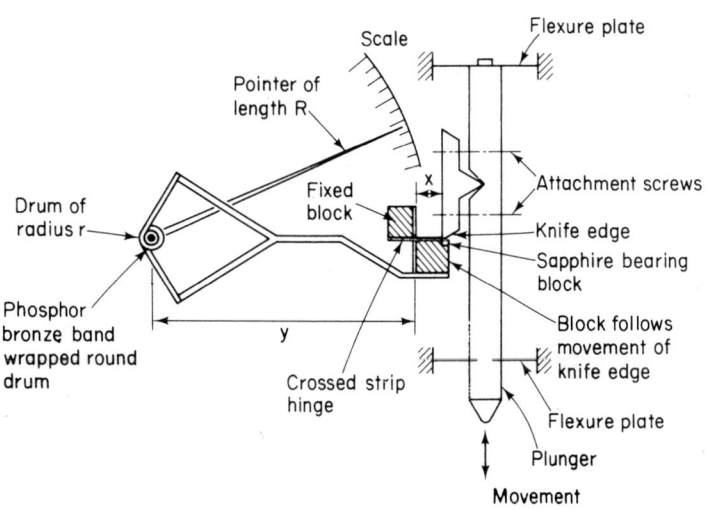

Figure 3.8 Sigma comparator

of the plunger causes the knife edge to move and so result in movement of the block at the end of the Y-shaped arm. The Y-shaped arm rotates about the hinge and a magnified movement is produced at the other end of the arm. This movement transmits rotatory movement to a drum by means of a phosphor bronze band wrapped round the drum. A pointer attached to the drum is then made to move across a scale.

OPTICAL LEVERS   A ray of light incident on a mirror at an angle of incidence *i* will be reflected with an angle of reflection equal to this angle of incidence (*Figure 3.9a*). If the mirror rotates through an angle θ then the angle of incidence will change to $(i + \theta)$. The angle of reflection must then also become $(i + \theta)$. Before the mirror rotation the angle between the incident and reflected rays was $2i$ and after the rotation $2(i + \theta)$. The angle between the incident and reflected rays has thus changed by 2θ

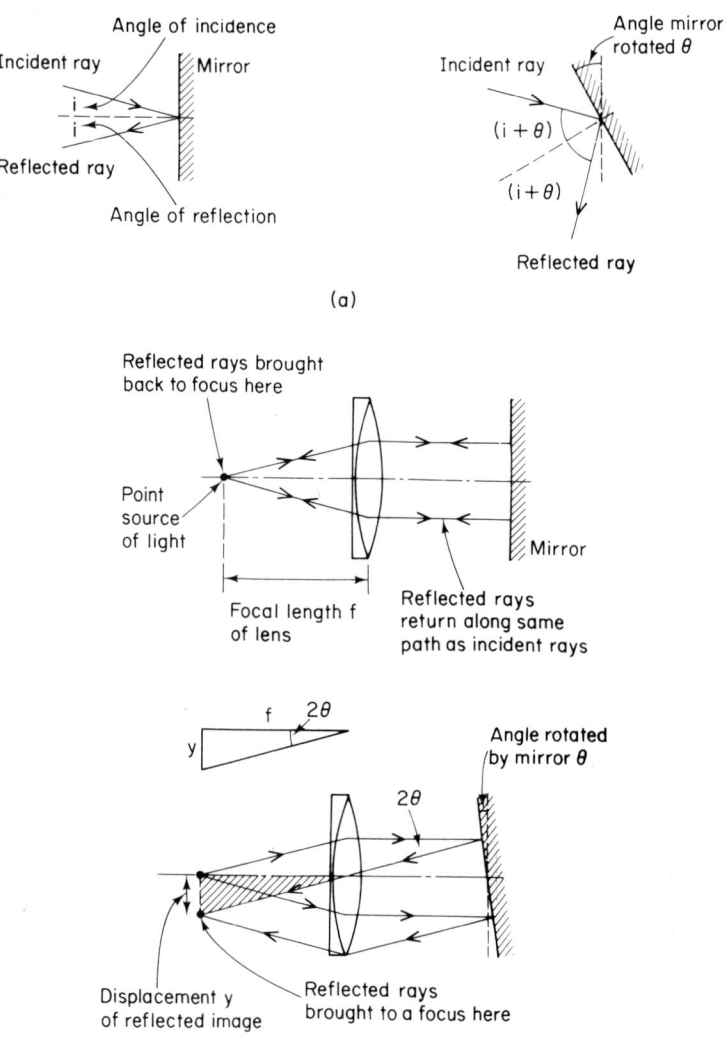

(a)

(b)

Figure 3.9   (a) Optical lever (b) Optical lever with a collimating lens

as a result of a rotation θ. If a mirror is rotated through an angle the reflected ray turns through twice that angle. This principle can be used to give a magnification, equal to two, of an angular displacement.

Figure 3.9b shows the same principle being used with a lens. A source of light placed at the focal point of a convex lens will result in parallel rays of light emerging from the lens. The lens is said to produce a *collimated* beam of light and is called a *collimating lens*. If parallel rays of light fall on to a plane mirror at right angles to the surface of that mirror, i.e. an angle of incidence of zero, then the reflected rays return along the same path as the incident rays. If however the parallel rays do not meet the mirror at zero angle of incidence then the reflected rays do not return along the same path as the incident rays. The result is as shown in *Figure 3.9b*. When the rays return along the same path through a convex lens then the image is formed at the same position as the point source from which the rays initially emerged. When they do not return along the same path then the image is formed at a different position to the initial point source. The displacement of this image from the initial position is a measure of the angle through which the plane mirror has rotated.

$$\frac{y}{f} = \tan 2\theta$$

If θ is small the expression can be simplified to

$$\frac{y}{f} = 2\theta$$

where f is in radians. Thus for a displacement y which is large for a given angular displacement θ a large focal length is required.

With a lens of focal length 100 mm and an angular displacement of the mirror of 1°, i.e. about 0.018 radians, then the displacement y of the image from the source will be

$$\frac{y}{100} = 2 \times 0.018$$

$$y = 3.6 \text{ mm}$$

If the focal length had been only 50 mm then the displacement would have been only 1.8 mm. A greater magnification is achieved with the longer focal length lens.

The distance between the reflector, i.e. the mirror, and the lens has no effect on the magnification, only the focal length determining the magnification. There is one other point—the angle through which the mirror rotates must not be so large that the beam of reflected light bypasses the lens altogether. This factor limits the distance between the mirror and the lens.

Figure 3.10 shows the *auto-collimator*, an instrument designed for the measurement of angles and based on the above principle. An illuminated target wire is mounted at the focus of the collimating lens. The target wire and the reflected image of the target wire are viewed through an eyepiece which incorporates a scale graduated in 0.5 minute intervals (a minute is 1/60th of a degree). In addition there is a pair of setting wires, the position of which can be adjusted by means of a micrometer graduated in 0.5 second intervals (a second is 1/60th of a minute). When the instrument is being used the setting wires are moved

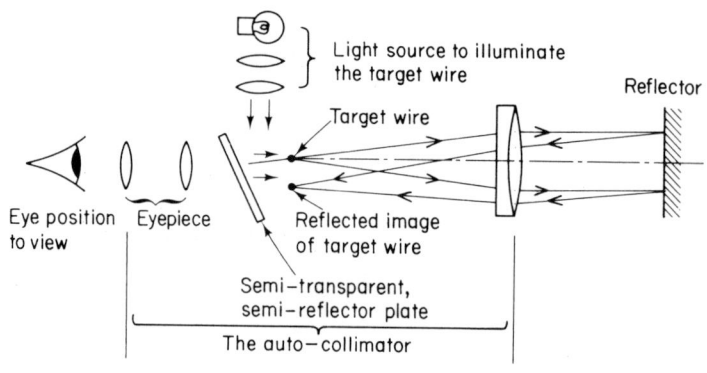

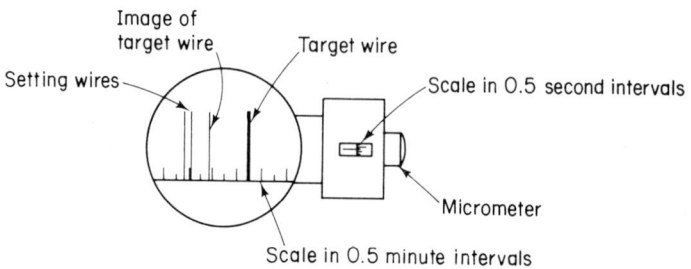

View through the eyepieces

Figure 3.10   The auto-collimator

so that they straddle the target wire image and the micrometer reading taken for this setting. The position of the target wire image is taken for the reflector in its initial state and after it has rotated, the difference between these two readings is related to the angle through which the reflector has rotated.

Figure 3.11 shows an example of the use of an autocollimator in the measurement of straightness. The reflector is mounted on a small angle bracket and readings are taken with the auto-collimator when the

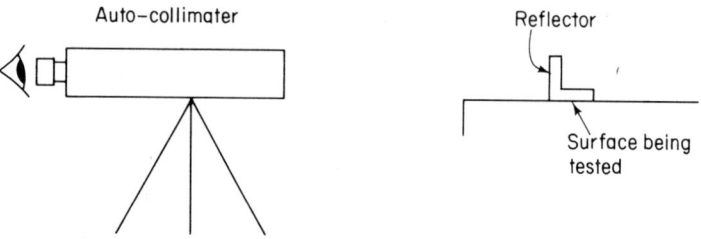

Figure 3.11   Straightness measurement with an auto-collimator

bracket is at different positions along the surface being tested. Any deviation from straightness of the surface along which the bracket is moved will result in a change of angle of the reflector and hence a change in reading for the auto-collimator.

Figure 3.12 shows another instrument, the *Angle Dekkor*, which depends on the same principle. A scale engraved on a screen is illuminated, the screen being in the focal plane of the collimating lens. The

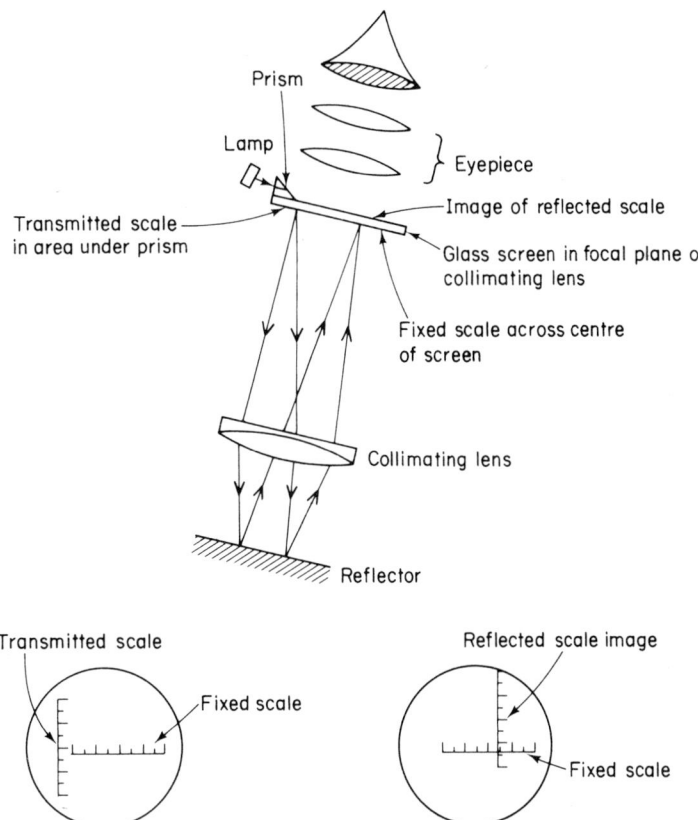

Figure 3.12   The Angle Dekkor

light reflected at the reflector is focused back on to the screen, in a position displaced from the original scale. The screen in the area where the reflected image is received has another scale, this scale being at right angles to the reflected image scale. The displacement of the reflected screen image above or below the datum line of the fixed scale gives a measure of the change in angle of the reflector in one direction, the movement of the reflected screen image along the fixed scale gives a measure of the change in angle of the reflector in a direction at right angles to the other direction. The scales are graduated in minutes and it is possible to estimate to about 0.2 minutes. The instrument is not as sensitive as the auto-collimator.

GEAR TRAINS   A *simple gear train* can be used as a signal conditioner. *Figure 3.13* shows a simple gear train used to give an output larger than the input. Rotation of the larger gear wheel results in rotation of the smaller gear wheel. Each tooth on the larger wheel fits into a corresponding space between teeth on the smaller wheel as the larger wheel rotates. Thus a movement resulting in one tooth on the larger wheel being replaced by the next tooth must result in a similar movement of the smaller wheel. Thus if there are 36 teeth on the larger wheel then one complete revolution of that wheel must mean that the smaller wheel must rotate so that 36 spaces between teeth move round. If the second wheel has

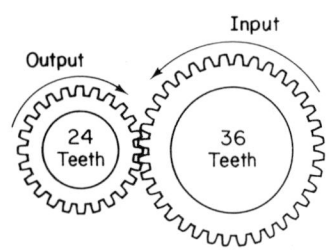

Figure 3.13

only 24 teeth then it would have to complete 36/24 = $1\frac{1}{2}$ revolutions for this to happen. Thus while the larger wheel rotates through one revolution the smaller wheel rotates through $1\frac{1}{2}$ revolutions. A pointer attached to the smaller wheel would thus have moved through $1\frac{1}{2}$ times the angle it would have if it had been attached to the larger wheel. There is a magnification of $1\frac{1}{2}$.

$$\text{Magnification} = \frac{\text{number of teeth on input gear wheel}}{\text{number of teeth on output gear wheel}}$$

The Bourdon tube pressure gauge shown in *Figure 1.2* uses a rack-and-pinion arrangement to give magnification. This is the same principle as the gear train referred to above.

Compound gear trains involving more gear wheels can be used to give greater magnifications. *Figure 3.14* shows such a compound gear train.

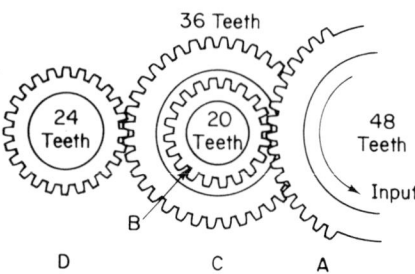

Figure 31.4

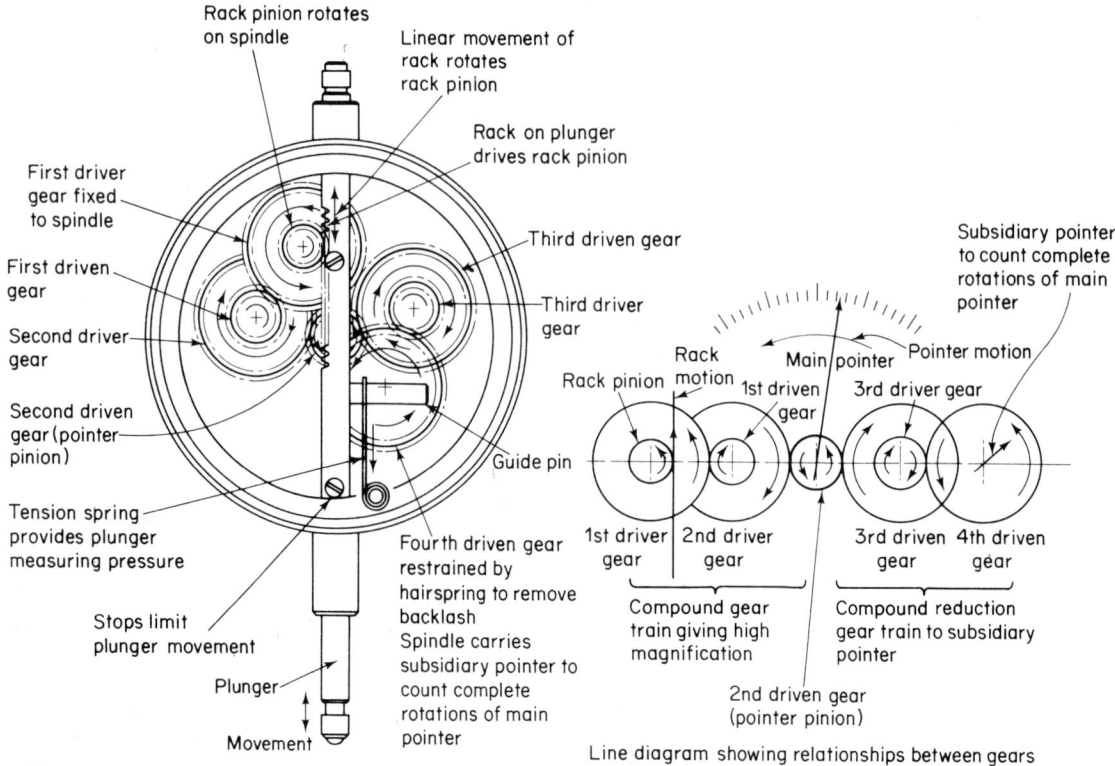

Figure 3.15 Dial indicator gauge

The input is via gear wheel A which has 48 teeth. The teeth on this wheel are in contact with the spaces between the teeth in gear wheel B which has 20 teeth. Thus one rotation of A results in 48/20 = 2.4 revolutions of B, i.e. a magnification of 2.4. Gear wheels B and C are on the same shaft so that one rotation of B means one rotation of C. Thus when A rotates by one revolution both B and C rotate by 2.4 revolutions. Gear wheel C has 36 teeth and drives gear wheel D which has 24 teeth. Thus one revolution of C results in 36/24 = 1.5 revolutions of D. Thus 2.4 revolutions of C mean 2.4 × 1.5 = 3.6 revolutions of D. The overall magnification of the gear train is thus 3.6.

$$\text{Magnification} = \frac{\text{teeth on A}}{\text{teeth on B}} \times \frac{\text{teeth on C}}{\text{teeth on D}}$$

Figure 3.15 shows the dial indicator gauge. A high magnification of the movement of a plunger is obtained by the use of a rack and pinion with gear train. The linear movement of the plunger is translated into a rotation by means of the rack and pinion. This rotation is then magnified by means of a gear train.

TWISTED STRIPS

Figure 3.16 shows the basic principles of the *Johansson extensometer.* The basis of the instrument is a twisted metal strip. Extension of the sample results in a movement of the 'free' knife edge which rotates and

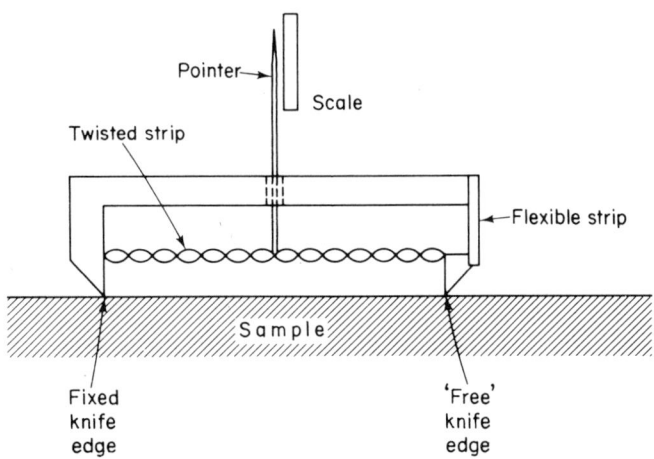

Figure 3.16 Johansson extensometer

results in the twisted strip partially unwinding. The untwisting of the strip causes a pointer to move across a scale. The movement of the pointer is considerably greater than the movement of the knife edge. Magnifications as great as 5000 are possible. The twisted metal strip is the signal conditioner.

WHEATSTONE BRIDGE

There are many transducers which have as their output a change in electrical resistance. The electrical resistance strain gauge and electrical resistance thermometer are examples of such transducers. If the resistance changes are quite large, as in a thermistor, then a simple circuit such as that shown in *Figure 3.2* might be adequate—particularly if the system is not required to give anything more than rough results. For many transducers the resistance changes are not very large. For example, a strain gauge with an initial resistance of 100 Ω and a gauge factor of 2.0 subject to a strain of 2×10^{-5} gives a resistance change of only 0.004 Ω. The circuit shown in *Figure 3.2* would not be suitable in such a case.

A circuit that is widely used for the measurement of small changes in resistance is the *Wheatstone bridge*. *Figure 3.17* shows the general arrangement of the bridge circuit. *P, Q, R* and *S* are resistors, *P* being the transducer. Between points *A* and *C* of the bridge there is a battery, between points *B* and *D* a sensitive galvanometer. In one way that the bridge is used the function of the galvanometer is only to determine whether or not there is a current.

For no current to flow through the galvanometer there must be no potential difference between points *B* and *D*, i.e. they must both be at the same potential. Thus the potential difference across *P* must be the same as that across *R* i.e. potential difference between *A* and *B* the same as that between *A* and *D*.

Thus $\quad I_1 P = I_2 R.$

I_1 and I_2 are the currents indicated on *Figure 3.17*, *P* and *R* are the resistances of resistors *P* and *R*.

Similarly the potential difference between *B* and *C* must be the same as that between *D* and *C*. As no current flows through the galvanometer the current through *Q* must be the same as that through *P*—there is no where else for it to flow. Similarly the current through *S* must be the same as that through *R*.

Hence $\quad I_1 Q = I_2 S$

If the above two equations are rearranged to eliminate the currents the result is

$$I_1 P = I_2 R = \frac{I_1 QR}{S}$$

$$P = \frac{QR}{S}$$

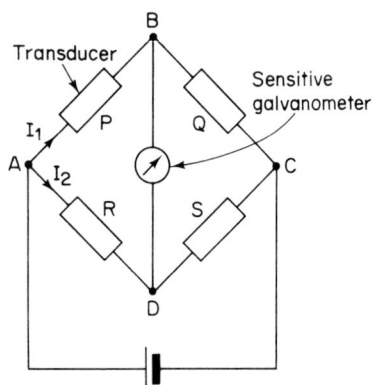

B

Transducer

Sensitive galvanometer

I_1

P

Q

A

I_2

C

R

S

D

Figure 3.17 The Wheatstone bridge

This is the relationship between the resistances when no current flows through the galvanometer. By making *R* and *S* fixed resistors and *Q* variable the resistance of the transducer *P* can be found, *Q* being adjusted until for a particular value of *P* there is no current. If the resistance of the transducer *P* changes then *Q* can be adjusted until again there is no current.

Suppose the resistance of *S* to be 1000 times that of *R*, i.e. *R/S* = 1/1000. When no current flows *Q* must have a resistance 1000 times that of *P*, i.e. P/Q = 1/1000. Thus if *P* is a strain gauge with a resistance of 100 Ω then *Q* must have a value of 100 000 Ω. If when the gauge is strained its resistance rises to 100.004 Ω, i.e. a change in resistance of 0.004 Ω, then the value of *Q* must be changed to 100 004 Ω, i.e. a resistance change of 4 Ω. This change in resistance is easily arranged by adjusting a resistance box.

The Wheatstone bridge is a very sensitive method of determining resistance. One of the reasons for this is that the galvanometer has only to detect whether or not there is a current, the size of the current being unimportant. Such a method is called a *null method*.

The Wheatstone bridge is often used in the unbalanced condition, with all the arms of the bridge generally being about the same resistance. The current indicated by the galvanometer is nearly proportional to the change in resistance of the transducer provided the change in resistance from the balance condition is small. The galvanometer reading can be calibrated by inserting known resistances in a bridge arm. The bridge used in this way has the advantage of being direct reading without the necessity of balancing the bridge by adjusting one of the resistors. It allows the use of the bridge with a chart recorder (see Chapter 4). The disadvantages of using the bridge in this way are that the accuracy of the null method is lost and the calibration depends on the voltage used to supply the bridge, unlike the null method. If the voltage changes, a new calibration is required.

There are many transducers where compensation has to be made for temperature changes. Thus in the case of the electrical resistance strain gauge the gauge responds not only to strain but also to temperature. For the strain results to be obtained the temperature effects have to be compensated for. This can be done, in the bridge shown in *Figure 3.17*, by having the transducer for the strain measurement as *P* and a compensating transducer as *R*. The two transducers should be identical, the only difference being that *P* is subject to both the strain and the temperature changes and *R* is subject to only the temperature changes. If *P* and *R* had the same resistance change the balance of the bridge would not be affected. The changes would in general only be the same if *P* were not subject to strain but only the same temperature change as *R*. If *P* is subject to strain then the bridge will go out of balance, but the bridge has only to be restored to balance by an amount related to the resistance change produced by the strain. The temperature effects are thus compensated for.

An alternative way of using a Wheatstone bridge with strain gauges which also compensates for temperature changes but which gives a greater sensitivity to strain is to arrange two gauges so that one is subject to tensile strain and the other to compressive strain. Thus they might be placed on opposite sides of a bending beam (*Figure 3.18*). If one of the gauges is used as *P* and the other as *R* in the bridge then

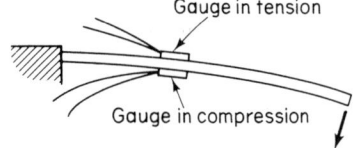

Figure 3.18 Cantilever with two strain gauges

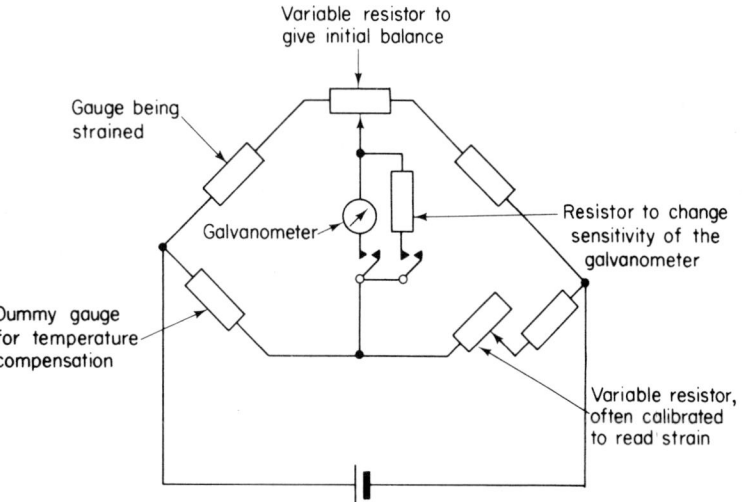

Figure 3.19 Practical form of Wheatstone bridge for use with strain gauges

temperature changes are compensated for. One of the gauges will however increase in resistance as a result of the strain and the other will decrease. The result will be a greater out of balance condition than if just one gauge had been used for the strain. Such an arrangement is commonly used in load cells (see Chapter 5) where the output is calibrated in terms of load.

In a practical form of Wheatstone bridge for use with electrical resistance strain gauges the ratio Q/S can be selected and the balance condition achieved by varying resistors, one of which is a slide wire. *Figure 3.19* shows the type of circuit used. The resistors in parallel with the galvanometer are to make the instrument less sensitive. Thus when the bridge is first being balanced the galvanometer should be at its least sensitive. When the bridge is balanced at this sensitivity then the sensitivity is increased and the balance position improved. This avoids damage to the galvanometer by it being used with too large a current.

THE POTENTIOMETER

Some transducers have an e.m.f. or potential difference as their output, others can have their outputs changed into potential differences. The potentiometer is one method that can be used to transform such outputs into signals which can be recorded or displayed. Transducer outputs that are currents can be transformed into potential differences by passing the currents through resistors and detecting the potential differences across the resistors.

The *potentiometer* is a null method based on the idea of opposing one potential difference with another so that they cancel and there is no resulting flow of current. *Figure 3.20* shows the basic potentiometer. A variable potential difference is produced between the end of a resistor and a slider moving along the resistor, a constant current flowing through the resistor. When this potential difference is applied in opposition to the potential difference being measured no current will flow between them when the two have the same value. The value of the potential difference at which this occurs is related to the distance the slider has moved along the resistor. Thus the potential difference is converted into a displacement which can be directly indicated.

The displacement scale can be calibrated in terms of potential difference by applying known potential differences to be measured by the instrument. The standard potential differences are generally obtained from specially designed cells which have accurately known e.m.f.s under specified operating conditions. If the resistor in the potentiometer has a constant resistance per millimetre of its length then the displacement L of the slider from one end will be directly proportional to the potential difference.

$V \propto L$ or $V = kL$, where k is a constant.

With the standard potential difference V_s and a displacement L_s,

$$V_s \propto L_s \text{ or } V_s = kL_s.$$

Hence
$$\frac{V}{V_s} = \frac{L}{L_s}$$

A standard cell often used is the *Weston cell* and this has an e.m.f. of 1.01876V at 15°C. If with a potentiometer the displacement L_s of the slider were 452 mm with such a cell, what would be the value of the unknown potential difference if the balance condition were achieved

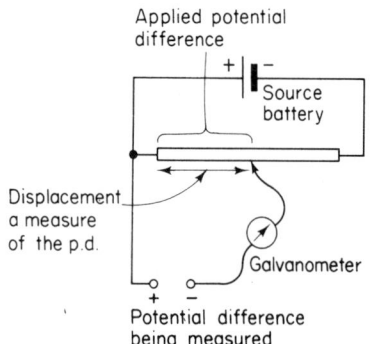

Applied potential difference

Source battery

Displacement a measure of the p.d.

Galvanometer

Potential difference being measured

Figure 3.20 The basic potentiometer

with the slider displacement of 356 mm? The balance condition is when no current is detected by the galvanometer.

$$\frac{V}{1.01876} = \frac{356}{452}$$

Hence to an accuracy of three figures, i.e. approximately the accuracy with which the slider displacement was measured, the potential difference is 0.802 V.

The accuracy of the potentiometer can be improved if the potential difference per millimetre slider displacement can be increased. In the above example there is 1.01876 V over 452 mm, i.e. 0.00225 V/mm. One millimetre is about the limit with which it is reasonably easy to set a slider. One way of increasing the voltage per millimetre is to increase the length of the resistor across which the slider moves and across which the source potential difference is applied. The length of the resistor can effectively be increased by putting another resistor in series with the resistance element across which the slider moves (*Figure 3.21*). Thus if the potentiometer has a resistor in series with it which is equivalent to 2000 mm of the resistance element, then the situation might be:

With standard cell 1.01876 V, balance with 2000 mm plus 634 mm. With unknown potential difference *V*, balance with 2000 mm plus 235 mm.

$$\frac{V}{1.01876} = \frac{2235}{2634}$$

$$V = 0.8644 \text{ V (correct to four figures)}$$

The potential difference per millimetre is about 0.0004 V/mm.

If the potential difference being measured is considerably less than that of the standard cell the arrangement shown in *Figure 3.21* for extending the length of the resistor across which the slider moves can still be used but the potential difference being measured can be applied just across the slider resistor (*Figure 3.22*). This is the situation with thermocouples, the e.m.f. often being microvolts per °C.

There are two commercial forms of potentiometer used in measuring systems, one in which the balance point is found by manually moving

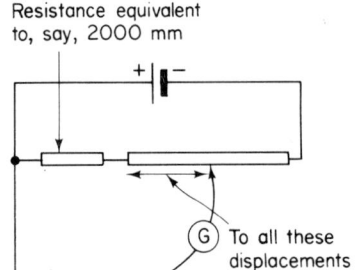

Resistance equivalent to, say, 2000 mm

To all these displacements add 2000 mm

Figure 3.21 Increasing the accuracy of a potentiometer

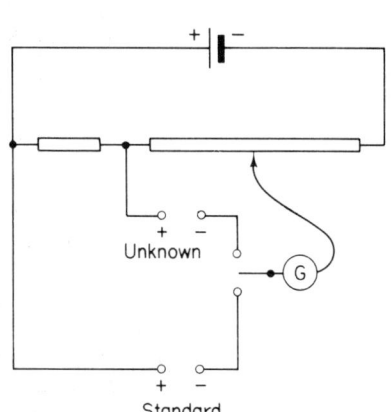

Unknown

Standard

Figure 3.22

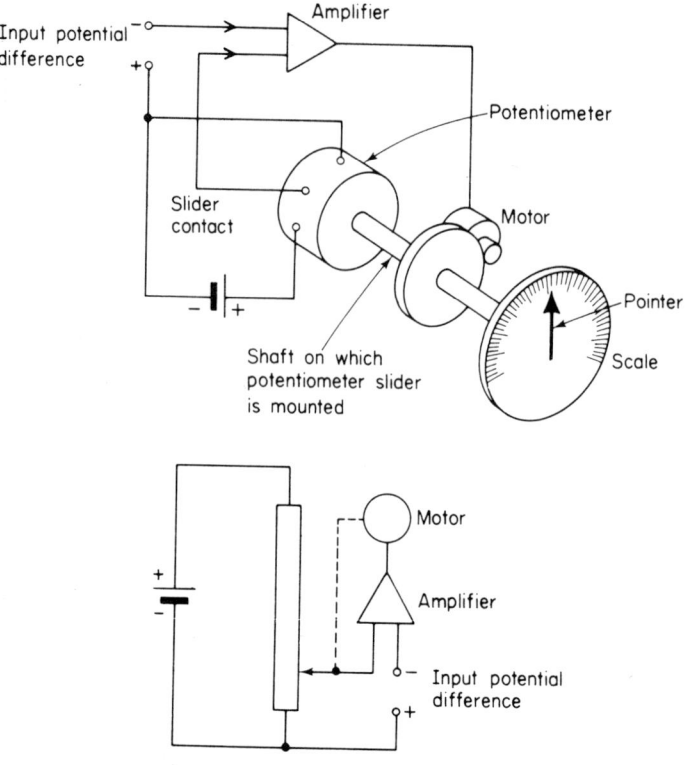

Figure 3.23 A self-balancing potentiometer

the slider across a resistor and the other which balances itself by the use of a motor. *Figure 3.23* shows the form of the *self-balancing potentiometer*. The difference between the potential difference being measured and that across the potentiometer resistor is fed via an amplifier to a motor. The motor shaft rotates and causes the potentiometer slider to move across the resistor until there is no difference between the two potential differences. When this happens the motor ceases to operate. The position of the slider at that point is then a measure of the potential difference. The potentiometer is generally in the form of a pot with the slider moving in a circular arc over the wire resistor. The position of the slider can thus be indicated by a pointer attached to the shaft to which the slider is attached.

Figure 4.13 shows how this principle of the self-balancing potentiometer is applied in a servo-recorder.

AMPLIFIERS   *Amplifiers* are used to increase the strength of electrical signals. A *voltage amplifier* can be used to make a voltage signal larger, i.e. the output is a voltage higher than the input voltage. A *current amplifier* can be used to enlarge a current signal, i.e. the output is a current higher than the input current. *A power amplifier* is used to give an output power higher than the input power. If an amplifier is not designated specifically for voltage, current or power it will in all probability be a voltage amplifier.

Amplifiers can be either d.c. or a.c. amplifiers. A *d.c. amplifier* is one where the input may be of a constant level, e.g. an unvarying voltage. An *a.c. amplifier* is one where the input must be alternating, e.g. an alternating voltage changing from positive to negative voltage perhaps fifty times a second. An a.c. amplifier will not operate with a d.c. input, however a d.c. amplifier will operate with an a.c. input.

In Chapter 2 reference was made to the problem associated with the use of the piezo-electric transducer where a strain applied to an appropriate crystal resulted in one face of the crystal acquiring a positive charge and the opposite face a negative charge. This charge separation means that there is a potential difference. If however a moving-coil voltmeter were connected to the crystal faces the charge would leak away through the voltmeter and thus a voltage reading would not be possible. Results can however be obtained by using what is called a *charge amplifier.* This produces a voltage output which is related to the charge at its input.

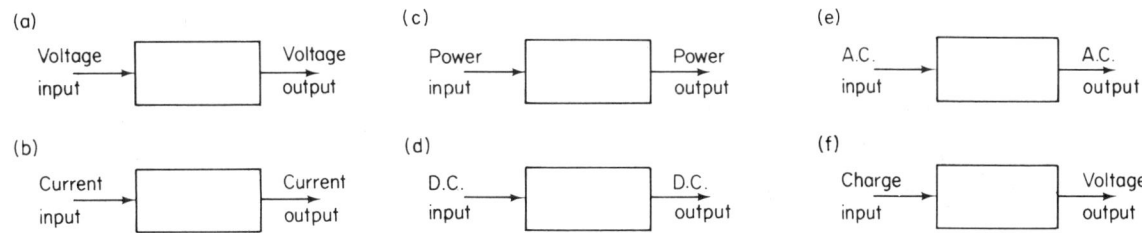

Figure 3.24 Amplifiers
(a) Voltage amplifier (b) Current amplifier (c) Power amplifier (d) D.C. amplifier (e) A.C. amplifier (f) Charge amplifier

Figure 3.24 shows the types of systems represented by the above forms of amplifier. All the amplifiers have common features—they have an input and an output and require electrical power. *Figure 3.25* is a specification for a charge amplifer.

PROBLEMS

(1)   State three examples of signal conditioner and describe how each of them conditions the signal.

(2)   A lever 100 mm long is to be used to magnify a displacement by a factor of 20. Where should the lever be pivoted when the input signal is at one end of the lever and the output signal at the other end?

(3)   How would you alter the design of the extensometer shown in *Figure 3.4* so that instead of giving a magnification of 6 it will give a magnification of 15?

(4)   *Figure 3.8* shows a diagram of the Sigma comparator. Explain how the magnification of the input displacement is achieved.

(5)   Explain how large magnifications are achieved with the Huggenberger extensometer (*Figure 3.7*).

(6)   Explain how a simple gear train can be used to give a signal conditioner which produces a magnification.

(7)   A Bourdon tube pressure gauge (*Figure 1.2*) uses a rack-and-pinion arrangement to give a magnification of 30. Explain how such a magnification can be achieved with this type of arrangement.

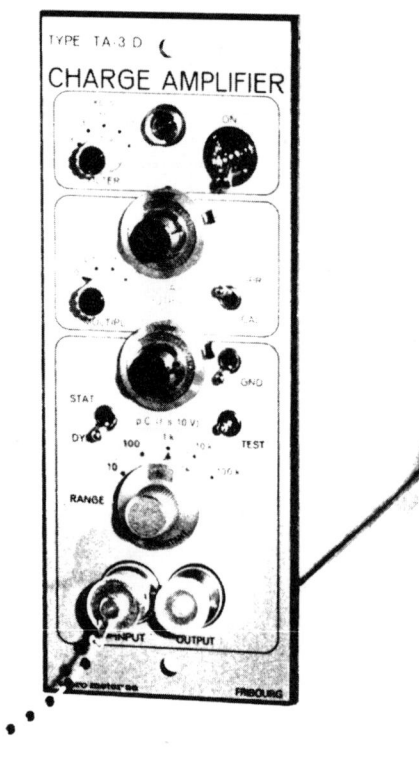

Figure 3.25 Charge amplifier, type TA-3/D (reproduced by permission of Vibro-meter Ltd)

GENERAL DESCRIPTION AND APPLICATION

A fully transistorized easy-to-operate charge amplifier for converting electrical charges into proportional voltages or currents. Many applications for laboratory measuring problems of static and dynamic phenomena in the fields of vibration-force-pressure.

MOSFET circuit with high loop gain, sensitivity not affected by cable length

Voltage and current output

Calibration of both voltage and current outputs by multiplier and digital potentiometer

Adjustable upper cut off frequency

Adjustable lower cut-off frequency for dynamic measurements

Remote Control of ground and test signals

Built-in test signal facilities

TECHNICAL SPECIFICATIONS

Input :
- Connection :   asymmetric
- Insulation : $> 10^{14}\ \Omega$
- Sensitivity :   10 pC to $1,1 \cdot 10^6$ pC for f.s.d.
- Adjustment :   continuously by combination of 10 turn potentiometer and 5-step multiplier
- Precision : $\pm$ 1% at 10^4 pC

grounding by build-in Reed-Relay, manually operated or remote controlled

Voltage Output :
- Connection :   asymmetric
- Voltage : $\pm$ 10V for full scale deflection
- Current :   max. $\pm$ 10 mA
- Internal Impedance :   22 Ω
- Linearity :   0,01% of full scale deflection
- Zero point :   adjustable

Current Output :
- Connection :   asymmetric
- Current :   max. 100 mA at full scale deflection
- Internal Impedance :   1 – 100 $\mu\Omega$, according to current adjustment
- Linearity :   0,1% of full scale deflection
- Frequency response :   3 db drop at 150 kHz
- Adjustment :   continuously by combination of 10 turn potentiometer and 4-step multiplier in the range 0 ÷ 100 mA

Temperature range : $- 10^\circ$C to $+ 60^\circ$C

Temperature influence :
- zero drift : $< 10^{-2}$ pC/$^\circ$C
- sensitivity variation : $< + 0,02\%/^\circ$C

Offset current of the charge amplifier :   max. 10^{-13}A $\hat{=}$ 0,1 pC/s

Open loop gain : $> 10^5$

Rise Time :   1 µs at position FILTER 300 kHz

Upper Cut-Off Frequency :   independent of the sensitivity adjust. chosen by 5 step low-pass filter : Cut-off (-3db): 1-3-10-30-300 kHz Descent : 12 db/octave

Influence of cable capacity :   0,5% sensitivity decrease per 100 m cable type C-6/. . . at multiplier position 10

Dimensions :   height/width/depth   177 x 70 x ca 295 mm

Weight :   ca. 2 kg (4,4 lbs)

Supply :   Mains 110-220-240 V 50-60 Hz

Ref. No. :   6050

(8) Explain how the Johansson comparator (*Figure 3.26*) operates.

(9) Explain how a Wheatstone bridge can be used (a) in the null condition and (b) as a direct reading instrument.

(10) Calculate the resistance needed to give a null reading in the following Wheatstone bridges. The resistors are labelled as in *Figure 3.17*.

	P	*Q*	*R*	*S*
(a)	50 Ω	100 Ω	?	100 Ω
(b)	?	50 Ω	50 Ω	100 Ω
(c)	?	200 Ω	200 Ω	400 Ω
(d)	100 Ω	20 Ω	40 Ω	?

(11) The Wheatstone bridge shown in *Figure 3.17* is used with an electrical resistance strain gauge subject to strain as P, a dummy gauge as R, a variable resistance box as Q and a fixed resistor of value 1000 Ω as S. P has a resistance of 120.1 Ω and R a resistance of 120.3 Ω. What is the value of Q needed for the null condition? By how much will the resistance of Q change if the strain gauge P is subject to a strain of 0.01% and the gauge has a gauge factor of 2.1?

(12) Explain how a dummy strain gauge can be used to compensate for room temperature changes during strain measurement with an electrical resistance strain gauge.

(13) Explain the principle of a potentiometer when used in the null condition.

(14) The following table gives the e.m.f. of a standard Weston cell at different temperatures. What is the effect on the results

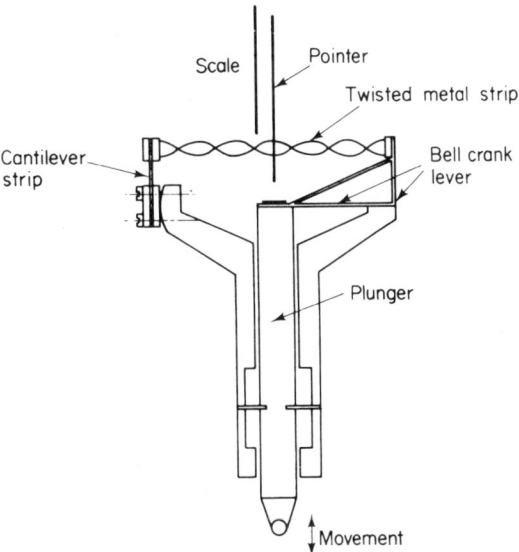

Figure 3.26 Johansson Mikrokator

obtained with a potentiometer if the 15°C e.m.f. value is used when the temperature may be 20°C or higher?

Temperature (°C)	10	15	20	25
E.m.f. (volts)	1.018 88	1.018 76	1.018 58	1.018 37

(15) Explain how a potentiometer can be used as a direct reading instrument.

(16) Identify the form of the input and output signals with the following amplifiers:
(a) D.C. amplifier.
(b) Current amplifier.
(c) Charge amplifier.
(d) Power amplifier.

(17) *Figure 3.25* gives the specification for a charge amplifier.
(a) What is the range over which it can be used for the measurement of charge?
(b) What is the accuracy?
(c) What outputs are possible?
(d) Over what temperature range can the instrument be used?
(e) What is the effect of temperature changes?
(f) What electrical supply is needed for the instrument?

ASSIGNMENTS

(1) Determine the magnification produced by a dial indicator gauge.

(2) Calibrate the out-of-balance current for a Wheatstone bridge using four identical strain gauges, in terms of strain.

(3) Design a potentiometer for the measurement of the e.m.f. of a copper/constantan thermocouple.

4 Display units

Aims: At the end of the chapter you should be able to:
Recognise the different forms of display unit and appreciate their different characteristics.
Select a display unit for a specified purpose.

WHAT ARE DISPLAY UNITS?

A *display unit* is a device which converts instrument signals from one form into another which is designed specifically for perception by a human observer. Thus in the case of a moving-coil meter (*Figure 4.1*) the movement of the coil causes a pointer to move across a scale. The pointer-scale system is the display unit designed to present the instrument signal to the human observer. In the case of a digital voltmeter (*Figure 4.2*) the output is presented in the form of a series of digits,

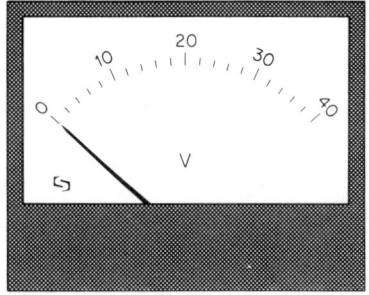

Figure 4.1 A moving-coil meter, an example of an analogue display

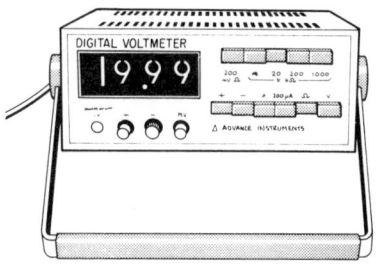

Figure 4.2 A digital voltmeter, an example of a digital display

i.e. numbers, on numerical indicator tubes. The display tubes present the instrument signal to the human observer. These two forms of display are called analogue and digital forms. An *analogue* display is one where the magnitude of the displacement of the pointer or indicator represents the size of the quantity being measured, i.e. the displacement is analogous to the value being measured. The movement of a pointer across a scale is an analogue display, the size of the movement being related to the size of the input to the display unit. With a *digital* display the output is a series of digits appearing on tubes, a screen or printed on a piece of paper.

Whatever form of display unit is used for a measuring system, consideration needs to be given to the interface between the system and the human observer, and thus the display should be designed to give the maximum operator efficiency. An important factor here is the ease with which the display can be read. A display designed for maximum operator efficiency is said to be *ergonomically designed.*

POINTER-SCALE INDICATORS

Figure 4.3 A stopwatch, an example of a pointer-scale display

Figure 4.3 shows one form of *pointer-scale indicator*. Such devices are indicators where a particular position of a pointer corresponds to a specific input signal. The scale across which the pointer moves can thus be calibrated in terms of the input. The pointer displacement is analogous to the size of the input signal. Pointer-scale indicators are thus, generally, analogue display units.

A pointer-scale indicator is a means of magnifying a small movement into something large enough to be determined easily by the movement of the tip of a pointer across a scale. The small movement is applied to the pivoted pointer a relatively short distance from the pivot (*Figure 4.4*). This results in a much larger movement of the end of the pointer, the amount of movement magnification being the ratio of the tip of pointer to pivot distance to the distance of the small movement application point from the pivot.

$$\text{Magnification} = \frac{\text{pointer tip to pivot distance}}{\text{small movement application point to pivot}}$$

Thus a large magnification can be achieved by applying the small move-

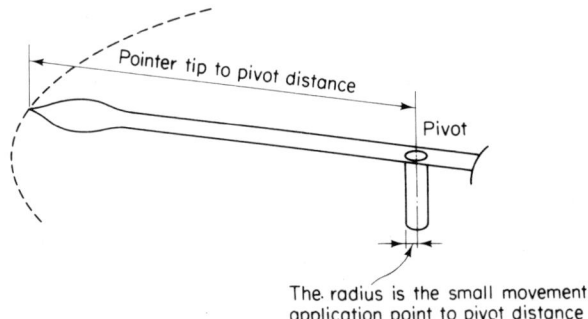

Figure 4.4

ment to a point close to the pivot and using a long pointer. The tip of the pointer can then give relatively large movements across a scale for quite small movements arising in the measuring system. The problem that arises from using a large pointer is that the mass of the pointer can become so large that the time taken for the pointer to respond to the input signal becomes quite large. The situation can be considered as a mass having to be accelerated, the larger the mass the smaller the acceleration for a given force and so the longer the time taken for it to cover a given distance. Depending on the pointer-scale arrangement used the response time may be of the order of a few tenths of a second or even seconds. Pointer-scale indicators do not respond very quickly to rapid changes in input.

The British Standards Institution specification for electrical pointer-scale displays states that the response should be such that four seconds, or more, after a relatively fast change in input signal which changed from zero to two-thirds of the full scale reading of the display the displacement of the pointer from the correct two-thirds reading should not be more than 1.5% of the full scale reading. Thus for an ammeter having a full scale reading of 100 mA, when the input changes from 0 to $\frac{2}{3} \times 100$ or $66\frac{2}{3}$ mA then four seconds or more after the change the pointer should be within 1.5 mA of the $66\frac{2}{3}$ mA reading.

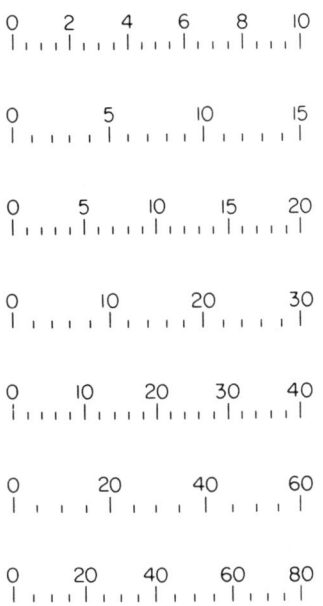

Figure 4.5 Some recommended scales

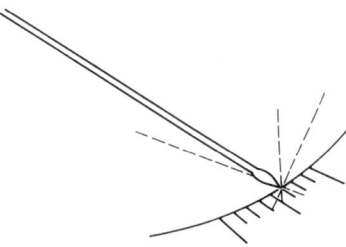

Figure 4.6 Parallax error. The reading depends on the angle at which the pointer is viewed

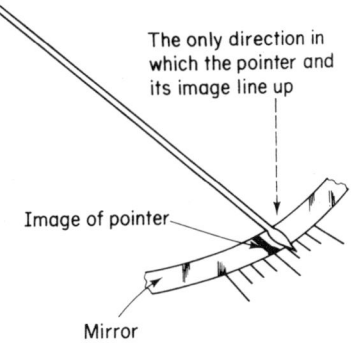

The only direction in which the pointer and its image line up

Image of pointer

Mirror

Figure 4.7 Using a mirror to eliminate parallax errors

A multimeter, AVO model 8 MK S, includes the following in its specification:

Response time: 1 second to full scale.

This means that when the instrument is used, on any of its scales, that it will take one second to reach a full scale reading when supplied with an input of that value.

An important consideration with the pointer-scale indicator is the design of the scale. British Standard BS 3693 'The design of scales and indexes' suggests that for rapid reading of a scale with the minimum observation error:

(1) the scale and figuring should be clearly laid out;
(2) there should be proper visual co-ordination between the pointer and the scale marks and between the scale marks and their associated numerals;
(3) the spacing of the scale marks and their thickness should be such that they are clearly legible at the maximum distance at which the scale is expected to be read.

Figure 4.5 shows some of the scale formats suggested in BS 3693. The maximum observation error that is considered acceptable is that in at least 90 out of every 100 observations a scale reading should not depart from the true value by more than ± 1% of the full scale reading when the maximum time for observation is two seconds and the scale is suitably illuminated.

Thus for an ammeter designed to this specification and having a full scale reading of 100 mA the maximum acceptable observation error is ± 1mA when the scale is viewed for at least two seconds.

An important consideration when scales are being chosen is the distance at which the observer views the scale. An approximate relationship between scale length L and the reading distance D for the ± 1% observation accuracy is

$$D = 14.4 L$$

Thus if a scale is to be read from a distance of 1 m then the scale length should be 0.069 m.

A moving-coil meter has a curved scale 127 mm long, designed according to BS 3693. Such an instrument could therefore be used at a reading distance of about 1.8 m for a ± 1% observation accuracy.

Errors can be introduced in the observation of the position of a pointer on a scale. In most pointer-scale indicator systems the pointer lies in a plane parallel to the scale but displaced a small distance from it. This is to allow the free movement of the pointer—it does not touch the scale. The reading observed will therefore depend on the angle at which the observer looks at the pointer in relation to the scale (*Figure 4.6*). Errors due to the pointer and scale being in parallel planes are called *parallax errors*. These errors can be minimised by having the distance between the two planes as small as possible. Another way of reducing the error is to mount a mirror close to the scale (*Figure 4.7*) and for the observer then to read the position of the pointer when the pointer and its own image are lined up. This ensures that the observer is viewing the pointer and scale in a direction at right angles to the mirror and hence the same viewing angle every time.

OPTICAL LEVER FORM OF INDICATOR

One of the problems with the pointer-scale indicator is the mass of the pointer. For a large magnification the pointer needs to be long but this results in an increase in mass and so a slower reacting indicator. This problem can be overcome by using a beam of light as the indicator. *Figure 4.8* shows the basic arrangement. The small movement being detected changes the angle a mirror makes with an incident beam of light. Such a change results in a movement of the beam of light reflected by the mirror. The reflected beam then moves across a scale and acts as a pointer. The distance of the scale from the mirror can be made quite large, much larger than the possible length of a pointer.

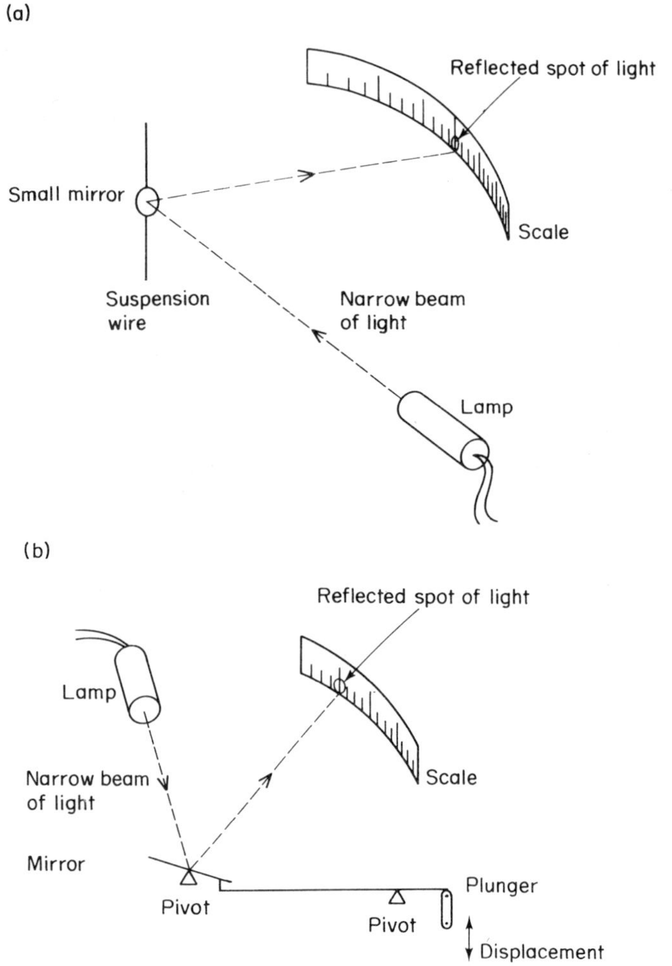

Figure 4.8 Optical levers
(a) The signal causes the suspension wire to twist and results in a movement of the spot of light across the scale
(b) Displacement of the plunger causes the mirror to rotate and results in a movement of the spot of light across the scale

If a beam of light makes an angle of incidence i with a mirror (*Figure 4.9*) then the angle of reflection will also be i, the angle of reflection always being equal to the angle of incidence. If the mirror then rotates through an angle θ, as shown in *Figure 4.9*, then the angle of incidence

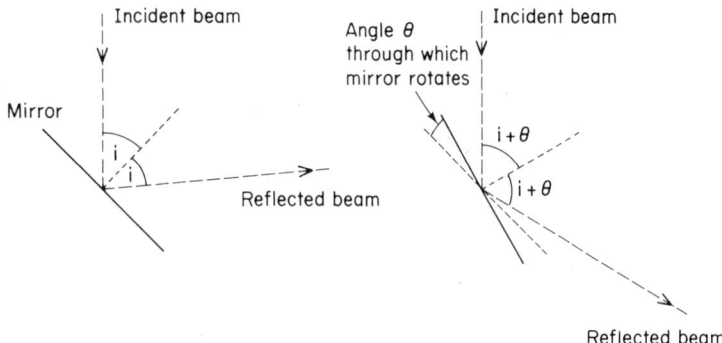

Figure 4.9 Reflection from a rotating mirror

changes from i to $(i + \theta)$ and so an angle of reflection of $(i + \theta)$. Before the mirror rotates the angle between the incident and reflected beams of light is $2i$; after the mirror rotates through an angle θ the angle between the incident and reflected beams is $2(i + \theta)$. The angle between the incident and reflected beams thus changes by 2θ when the mirror rotates through an angle θ.

CHART RECORDERS

Chart recorders are devices designed to produce graphical results directly on a sheet of paper. They show how one variable varies with respect to another. In many recorders they show how the input signal varies with time. They produce an analogue display. The input signal is transformed into a position on a piece of paper, i.e. the chart, and a marking mechanism leaves a record on the chart of that input.

Figure 4.10 shows one form of chart recorder. This is based on the use of the movement of a moving-coil galvanometer with the tip of the pointer carrying a marker. The deflection of the pointer resulting from the current through the galvanometer movement causes a mark to be made on the chart, the size of the displacement of the mark from the

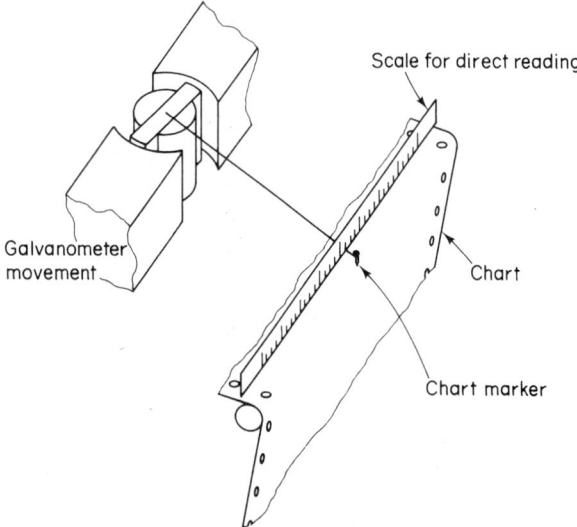

Figure 4.10 Galvanometer recorder movement

Galvanometer movement

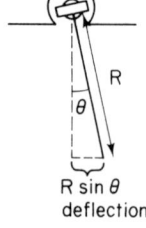

θ	θ radians	$\sin \theta$
0	0	0
5°	0.087	0.087
10°	0.175	0.174
15°	0.262	0.259
20°	0.349	0.342
25°	0.436	0.421

Figure 4.11

zero position being an indication of the size of the current through the galvanometer movement. The chart is driven by a clock motor, a sprocket wheel engaging in holes along the edges of the chart paper, and as the chart moves at a constant speed the displacement of the chart is proportional to time. A trace on the chart thus gives a graph of the galvanometer input variation with time.

The movement of the pointer of a galvanometer is in an arc, so for a marker attached to the end of such a pointer it is usual to employ charts which have curved grid lines. However, if the deflection of the pointer is small the chart can have straight grid lines. For small deflections $\sin \theta$ approximates to θ and thus the deflection $R \sin \theta$ approximates to $R\theta$ (see *Figure 4.11*). Some recorders use mechanisms to convert the rotational movement of the pointer to a straight line motion.

Another problem associated with the use of a galvanometer movement is the uniformity of the scale. It is convenient to have equal spacings between the grid lines to represent equal changes in the input. This is a reasonable approximation with the galvanometer movement for small angular deflections. Thus both straight line grids and equally spaced grid lines can be reasonably approximated with a moving-coil galvanometer movement restricted to small angular movements.

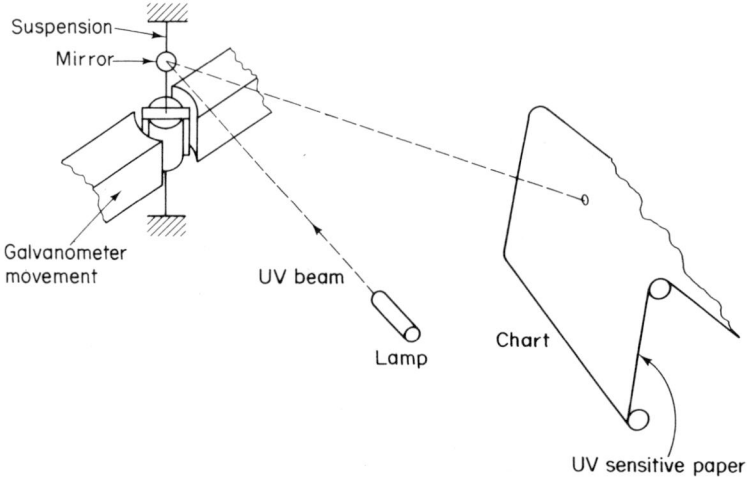

Figure 4.12 Mirror galvanometer movement

The marking mechanisms used with charts are generally pens, often felt or fibre tipped pens which can be removed when spent, or built-in pens using an ink reservoir which can be topped up, or an inkless method of marking. These latter mechanisms require specially coated papers. One form uses a beam of ultraviolet radiation instead of a

(a)

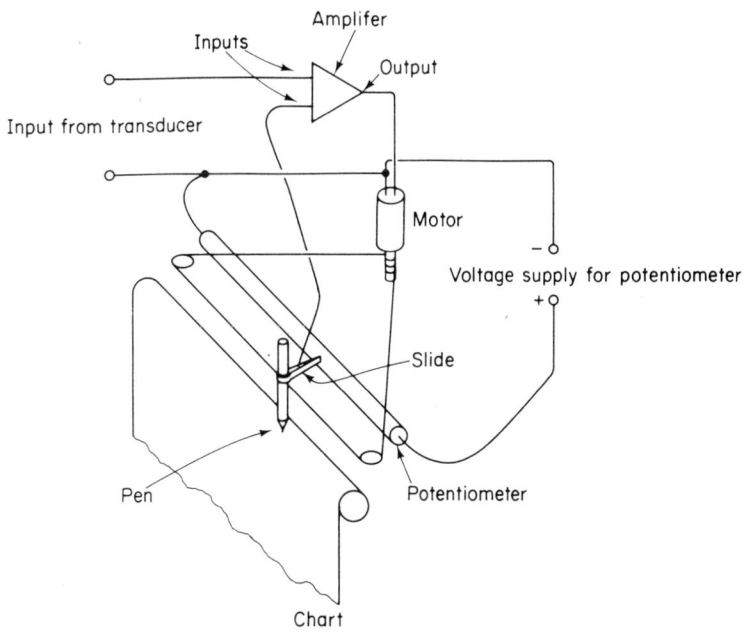

(b)

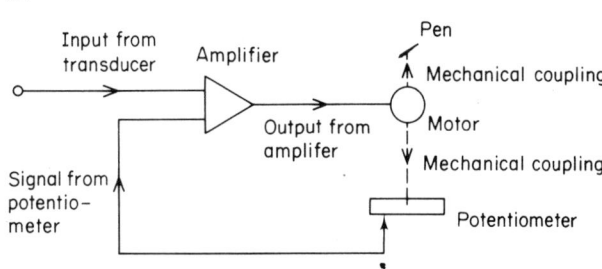

Figure 4.13 Servo-recorder
(a) Outline of a servo-recorder using a linear potentiometer
(b) The closed loop in the servo-recorder

pointer and an ultraviolet sensitive paper. *Figure 4.12* shows the basic arrangement of such a *UV recorder.*

Figure 4.13 shows another form of recorder, an example of an instrument known as the *servo-recorder.* This is not based on a galvano-meter movement but on directly driving a pen with a motor. When there is an input to the motor, the motor shaft rotates and so causes the pen to move across the chart. As long as there is an input to the motor it keeps on rotating and the pen keeps on moving across the chart. The movement of the pen causes a slider to move along a linear potentiometer which results in a change in the potential difference output from the potentiometer. The output is fed into an amplifier which also receives the signal input to the recorder. The two inputs to the amplifier result in an output only when the two inputs are different, and only then is the motor driven causing the pen to move across the chart. Thus the sequence could be as follows:

(1) Input from the transducer.
(2) The input to the amplifier results in an output.
(3) The output causes the motor to operate.
(4) The pen moves across the chart.
(5) As the pen moves across the chart the slider also moves over the potentiometer and so produces another input to the amplifier.
(6) The amplifier now has two inputs—the input from the transducer and that from the potentiometer. These initially partly cancel each other out. The pen keeps on moving and so increasing the input from the potentiometer until the two inputs to the amplifier completely cancel each other. When this happens there is no output from the amplifier and so no further movement of the pen.

The servo-recorder described above is known as a *closed-loop servo recorder*. This is because the signals follow a closed-loop path. The input from the transducer to the amplifier gives a signal to the motor and thence the potentiometer and so back to the amplifier again to modify its output.

On an average servo-recorder it may take the pen about 0.4 s to reach a full-scale deflection. This is much slower than the moving-coil meter which may take about 10 ms to reach the same deflection. The suspension wire form of meter movement, as in *Figure 4.12*, responds even more quickly. The amplifier in a servo recorder has to supply sufficient current to operate the motor. Because of friction there is a minimum current below which the current is too small to operate the motor. There is thus some error due to the recorder not indicating a small input from the transducer. This error is known as the *dead band*.

The following is part of the specification of a servo recorder.

Chart: width 120 mm, length 15 m.

Marking system: printing from a coloured inked ribbon.

Range: 0 to 5 mV.

Chart drive: electrical giving 5, 10, 20, 30, 60 and 120 mm per hour.

Accuracy: ± 1% of span.

Dead band: ± 0.3% of span.

Maximum response time: 2 s.

Thus in the above case the dead band is ± 0.3% of the span or ± 0.015 mV. The accuracy is ± 1% or ± 0.05 mV.

X-Y RECORDER   Some measuring systems have as output one variable, the y quantity, which varies with another variable, the x quantity. The output from such a system requires a recorder which will show how y varies with x. The servo recorder can be used for this purpose if the pen is driven by two separate motors, one motor controlling the movement of the pen in the x direction and the other the movement in the y direction. *Figure 4.14* shows an example of such a recorder.

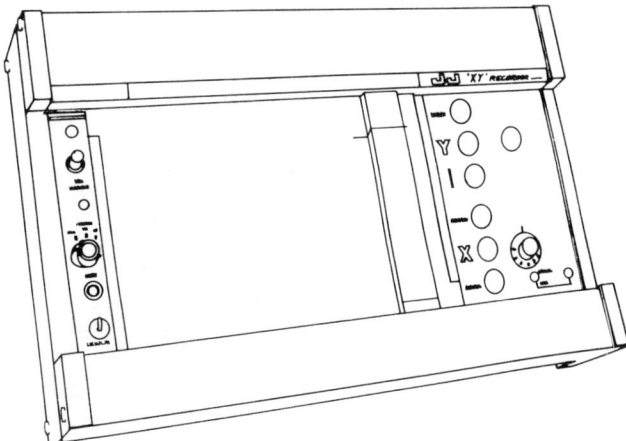

Figure 4.14 XY/t recorder (reproduced by permission of J. J. Lloyd Instruments Ltd)

CATHODE RAY OSCILLOSCOPE

The cathode ray oscilloscope gives a display in which one variable, y, is shown varying against another, x. In many uses of the cathode ray oscilloscope the x variable is chosen to be time.

The *cathode ray oscilloscope* consists essentially of a tube along which passes a beam of electrons. At the display end of the tube the beam strikes a fluorescent screen giving a small patch of light. The beam can be made to deflect in the y or x directions by potential differences applied to two sets of plates between which the beam passes. *Figure 4.15a* shows the general arrangement and *Figure 4.15b* a typical cathode ray oscilloscope. Internal signals can be supplied to the x pair of plates to give a signal which changes at a constant rate with time and so results in the beam of electrons being swept across the face of the fluorescent screen at a steady rate. Any input to the y plates which varies with time will then be marked out on the screen by the moving beam of electrons.

The cathode ray oscilloscope can be used to show how one variable varies with respect to another when the x and the y inputs are supplied with external signals.

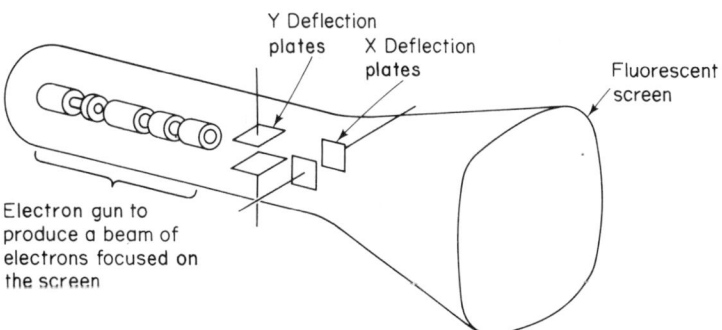

Figure 4.15a Cathode ray tube–internal electrode arrangement

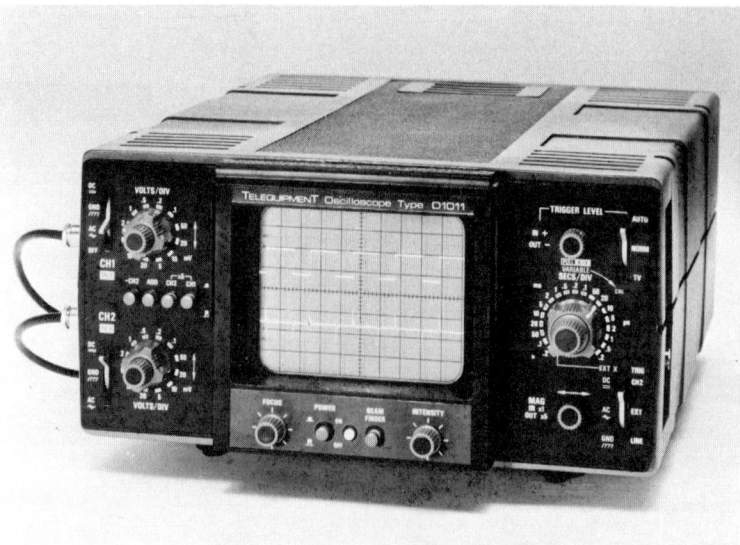

Figure 4.15b Typical cathode ray oscilloscope (Telequipment oscilloscope reproduced by permission of Tektronix (UK) Ltd)

A cathode ray oscilloscope can have a camera attached so that a record can be made of the signals portrayed on the screen. In this form the oscilloscope is rather similar to a chart recorder but one with a very fast response. Where a chart recorder would not be suitable for recording a very quickly changing signal, because of the long response time of the recorder, a cathode ray oscilloscope can be used. The response time of an oscilloscope is typically of the order of nano seconds (10^{-9} s).

DIGITAL DISPLAY   With a digital display the output from the transducer is displayed as a series of digits on tubes, a screen or printed on a piece of paper.

Figure 4.2 showed a digital voltmeter in which the voltage was displayed as a series of digits. There are a number of ways of presenting such displays (*Figure 4.16*).

(1) Numerical indicator tubes.
(2) Bar tubes.
(3) Semiconductor light-emitting diodes (LED).
(4) Liquid crystals.

With the *numerical indicator tube* the display occurs in a gas-filled tube having ten cathodes shaped in the form of numbers and an anode. When a sufficiently high potential difference is applied between the anode and the cathode a glow appears around the cathode. It is this glow which gives the display.

Bar tube displays comprise a number of straight light emitting tubes, i.e. the bars forming the numbers. Such tubes can be tiny fluorescent tubes, similar to those used for lighting, or minute electric lamps with the filament in the form of a bar rather than coiled as in the conventional electric lamp.

A display using *semiconductor light-emitting diodes* comprises a series of dots, each dot being one of the diodes. The numbers are made up by passing a current through the appropriate sequence of diodes

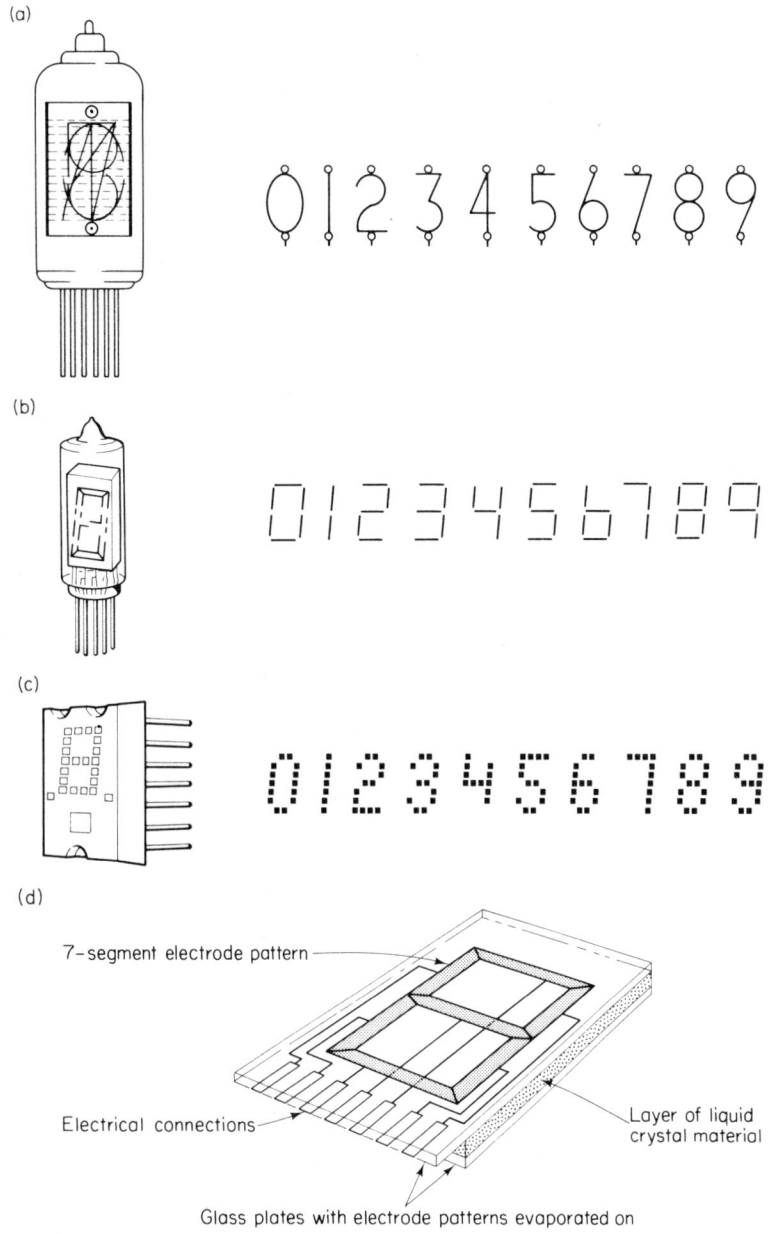

Figure 4.16 Typical forms of digital display units
(a) Numeral indicator tube   Typical form of numbers
(b) Bar tube   Typical form of numbers
(c) Semiconductor light-emitting diodes   Typical form of numbers
(d) Liquid crystal   Number form as (b)

causing them to light. The diodes are generally made of gallium phosphide or gallium arsenide phosphide to which small controlled amounts of impurities have been added. The colour of the light emitted by a diode depends on the material used for the diode and the type of impurity. A diode is a device which has a higher resistance to current flow through it in one direction than in the opposite direction. Light is emitted from these diodes when they are connected in such a way as to offer a current a low resistance path.

A *liquid crystal* display consists of a film of liquid crystal sandwiched between two transparent electrodes. The electrodes are thin layers of tin oxide on glass plates, the layer being in the form of the required number segments. When a potential difference is applied between the electrodes the liquid crystal between the electrodes becomes able to scatter light which previously would have just passed straight through, so that if light is directed on to the crystal from the front or the side, the area between the electrodes with the applied potential difference becomes bright.

An important aspect of such displays is the time taken for one number to change to another number, i.e. the time taken for one number to decay until no longer visible and the time taken for another number to become bright enough to be read. Lamps with filaments are quite slow in responding to change, one white hot filament having to cool sufficiently to no longer emit light and another having to warm up sufficiently to emit light. Semiconductor light-emitting diodes can respond very much more quickly and typically have response times of the order of 50 nanoseconds (50×10^{-9} s). Liquid crystals are much slower in responding than the diodes and typically have response times of the order of 50 milliseconds (50×10^{-3} s).

Specifications for digital display units might use the term 'bit'. A bit is the smallest piece of information that is registered and is generally one entry for a display unit. Thus for a display unit having a maximum capacity of 9999 bits the unit will show entries between 0 and 9 in four columns. This could be 0000 to 9999 or perhaps 0000 to 9.999.

PROBLEMS

(1)   A watch is a measuring system for time. How do the displays differ for analogue and digital watches?

(2)   State three forms of measuring system in which pointer-scale indicators are used for the display.

(3)   State an advantage and a disadvantage of making the pointer of a moving-coil meter very long.

(4)   A scale designed to BS 3693 has a full scale reading of 250 mA. What should be the maximum observation error, expressed in mA, for such a scale?

(5)   An instrument scale is to be read at a distance of 500 mm. What should be the length of the scale for a $\pm 1\%$ observation accuracy?

(6)   Explain what is meant by parallax error.

(7)   Design a straight scale to read 0 to 10 in halves and be read at a distance of 1 m.

(8)   A small mirror is mounted on a wire under tension. A beam of light is directed on to the mirror and the reflected beam observed on a scale (as in *Figure 4.8a*). Explain how you would calculate the angle through which the wire twisted from an observation of the movement of the reflected light beam on the scale.

(9)   *Figure 4.17* is taken from a manufacturer's catalogue.
(a) Which of the three indicators is capable of the least accuracy?
(b) Which indicator would be advisable with a measuring system having an accuracy of $\pm 0.1\%$?
(c) What is the relationship between the length of the scale and the maximum number of subdivisions?

Type	Scale Diameter	Length	Max Sub divisions	Inherent accuracy	Suitable for system accuracies of about—
Miniweigh	—	70 mm	50	± 0.75%	± 1%
GP4-10	257 mm	700 mm	500	± 0.1%	± 0.2%
PP4-20	445 mm	1280 mm	1,000	± 0.05%	± 0.1%

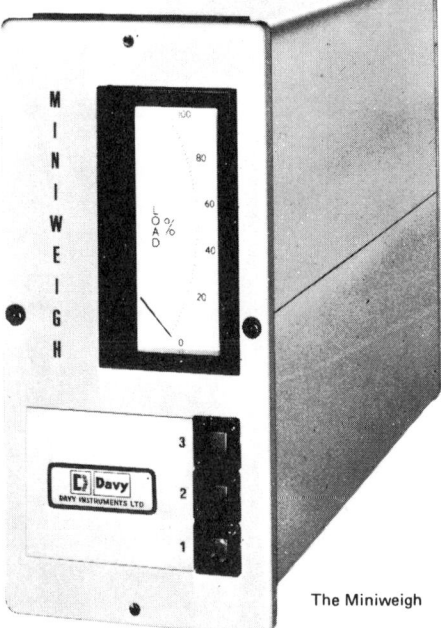

The Miniweigh

GP4-10 Indicator

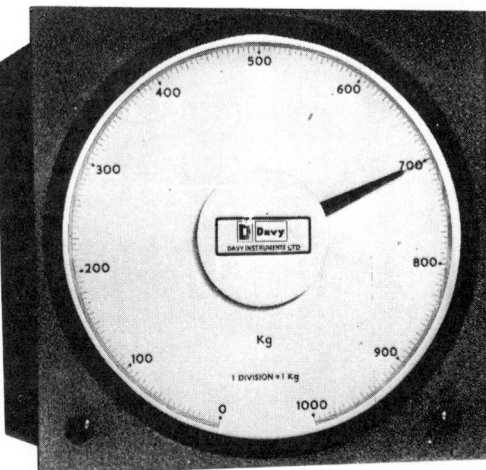

PP4-20 Indicator

Figure 4.17 Analogue indicators (reproduced by permission of Davy Instruments Ltd

(10) A chart recorder is said to be as follows:
Moving coil movement; full scale deflection 1 mA; 6 chart speeds of 20 to 5400 mm/h; siphon pen with separate ink reservoir; linkage system converts curvilinear pen movement into straight line trace.
(a) What is meant by 'moving-coil movement'?
(b) What is the maximum current input?
(c) With a chart speed of 20 mm/h how far apart on the chart will be two events that took place 30 minutes apart? What would be the separation of the two events if the chart speed were 5400 mm/h?

Figure 4.18   (reproduced by permission of J. J. Lloyd Instruments Ltd)

(d) Why is the linkage system used with the pen?
   (11) A chart recorder is said to be as follows:
Servo; 2 chart speeds 10 and 5 mm min^{-1}; input range 10 mV to 10 V calibrated; disposable nylon pen/ink cartridge.

(a) Explain the significance of the term 'servo' in this context.

(b) How far apart on the chart at each speed will be two events 2 minutes apart?

(12) A chart recorder is said to have a dead band of ± 0.4% of the maximum span. The maximum span of the recorder occurs with 10 mV. What is the significance of this dead band?

(13) *Figure 4.18* shows an advertisement for chart recorders. Describe in your own words the capabilities of the recorders. Where technical terms occur explain them—assume your report is to be read by non-technical staff.

(14) A digital indicator is specified as follows:

Maximum number of bits 999

Inherent accuracy ± 1 bit

Suitable for system accuracies of ± 0.1%

(a) What is meant by the term 'bit'?

(b) What is the inherent accuracy in terms of 'percentage of maximum reading'?

(15) Which types of display would be worth considering if you needed a display unit which could respond to an input which varied rapidly with time?

ASSIGNMENTS

(1) Calibrate an indicator, e.g. a chart recorder, a meter, a cathode ray oscilloscope. Explain how to set up the standards used for the calibration.

(2) Make a critical analysis of the readability of the different scales used on instruments. Design an investigation to determine their readability.

(3) Make an analysis of the readability requirements of a scale used in some measurement system in industry.

5 Measurement systems

This chapter consists of a number of sections devoted to different forms of measuring system. More systems are included than are specified in the TEC standard unit 'Engineering instrumentation and control IV'. Each section includes both problems and assignments.

Aims: At the end of each section you should be able to:
Explain and compare the different types of system available for the measurement concerned.
Specify the most likely form of system for a particular measurement.

The assignments have the following aims, though not all assignments will have all of the aims:
Analyse the requirements for the measuring system.
Specify a possible system.
Test and calibrate the system.
Write a report on the choice, test and calibration.

1. MEASUREMENT OF TIME AND FREQUENCY

The basic unit of time is the *second* (see Chapter 1 for definition). The basic unit of frequency is the *hertz* (Hz). An oscillator operating at a frequency of 1 Hz completes one cycle per second. Thus a frequency of 50 Hz means 1/50 s per cycle.

Time is measured in terms of some form of oscillation which may be that of a pendulum or a balance wheel in a watch. It could also be an electrical oscillation or even an atomic oscillation.

Time measurement systems can be either mechanically or electrically operated. Thus the *stopwatch* shown in *Figure 4.3* is a mechanically operated time measurement system. The watch is started by the mechanical operation of pressing the button at the top of the watch and stopped by again pressing the button. Thus to measure the time taken for some event requires the observer to see the event starting, press the button to start the watch, and then when the event is over to see this happening and again press the button to stop the watch. Time is taken for the observer to see the event and then react and press the button. This time is known as the *reaction time* of the observer. Typically it is of the order of half a second. Thus the accuracy of the timing operation is determined not only by the accuracy of the measuring system but also by the reaction time of the operator. For time intervals less than about 20 s the reaction time of the operator can have a significant effect on the percentage accuracy of the measurement. It is possible to avoid the reaction time inaccuracy of the operator with a mechanically operated system by using mechanical trip mechanisms to start and stop the timer. It is possible to reduce the effect of reaction time on an operator controlled system by using the same operator operating in the same conditions for starting and stopping the system. If the reaction time were the same for both starting and stopping then it would have no effect on the accuracy. This is not likely to occur

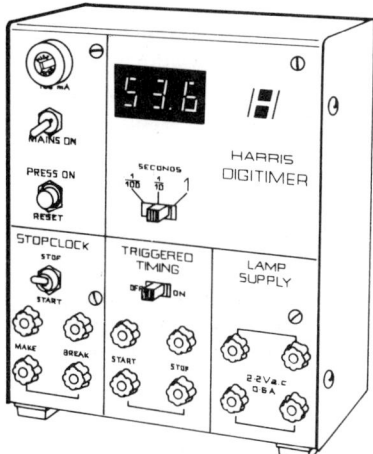

Figure 5.1 Electronic timer (reproduced by permission of Philip Harris Ltd)

with a high degree of accuracy but a trained operator might reduce the effect of reaction time on the overall accuracy. For small time intervals or for accurate measurement of time intervals an electrically operated timer is to be preferred.

Figure 5.1 shows a simple *electrical timer* having a digital display giving three figures. The display can have ranges of 1/100 s, 1/10 s or 1 s. Thus with the display switched to the 1/100 s range the maximum display will be 9.99 s, on the 1 s range the maximum will be 999 s. The timer can be started by making an electrical circuit and stopped by breaking an electrical circuit. Electrical circuits can be made and broken by mechanically operated switches, the operation being timed being itself the operator of the switches. When a circuit is 'made' the current rises from zero to some value capable of operating the timer. It is possible to use light-dependent resistors (see Chapter 2) to give the same type of result. When no light falls on the resistor it has a very high resistance and so the circuit in which that resistor is included passes a very small current, virtually zero and so is not capable of operating the timer. When light falls on the resistor its resistance drops to a low value, the current rises and so operates the timer. Thus the timer can be made to operate for the time interval during which light falls on the resistor. It is also possible to arrange for a timer to operate for the time interval during which there is no light falling on a light-dependent resistor. This method enables time intervals to be measured for, say, an object to move through a light beam (*Figure 5.2*).

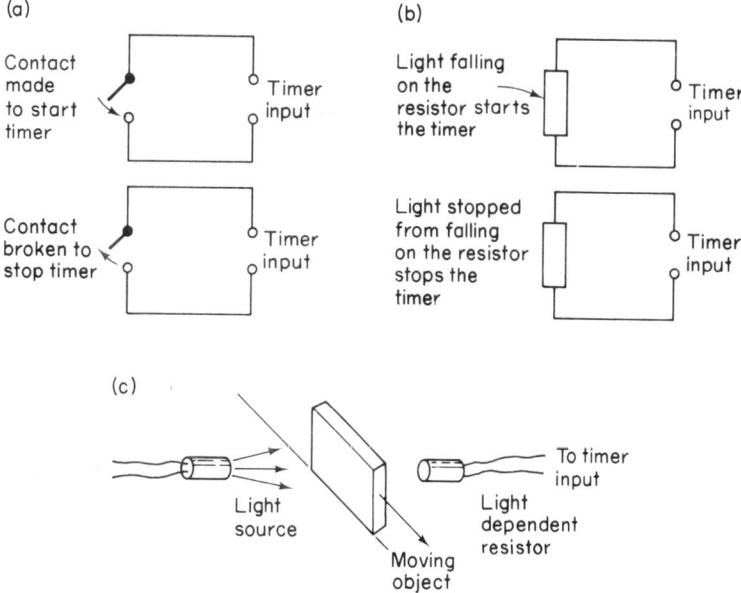

Figure 5.2 Measuring time intervals
(a) Mechanical starting and stopping of timer
(b) Light-controlled starting and stopping of timer
(c) Breaking the light beam starts the timer, the light beam falling on the resistor stops the timer

Electric timers generally base their time indication on the counting, in the timer, of electrical oscillations which can be derived from the 50 Hz mains oscillations. Each oscillation lasts 1/50 s and so a count of these enables times to be measured to 1/50 s. Timers used for time measurements smaller than this operate with higher frequencies, sometimes produced by an electrical circuit which increases the mains frequency by a constant factor, say a doubling to give 100 Hz and so measurement to 1/100 s. An electric timer is thus essentially a *pulse counter*.

The frequency of the electrical mains supply in the UK is 50 Hz and there is a statutory requirement that this should be maintained within ± 1%, thus constituting a limit on the accuracy of timers based on the mains frequency. Electronic oscillators give other frequencies, and can be more accurate than the mains frequency as a time or frequency standard when *stabilised*, generally by a quartz crystal. Their frequency is designed to line up with the frequency at which the crystal will mechanically oscillate.

Table 5.1 Typical example of part of the specification of an electronic analogue timer.

Five millisecond ranges 0–4, 0–10, 0–40, 0–100 and 0–400, and three seconds ranges, 0–1, 0–4 and 0–10 seconds; accuracy ± 1% full scale deflection on all ranges except the two shortest, on these ranges accuracy is ± 2% full scale deflection; input signals 3–300 V negative or 'negative-going', positive signals ignored. For operation on 100–120 V, 200–250 V, 40–60 Hz a.c.

PROBLEMS

(1.1) Explain how variations in the reaction time of an operator can affect the results obtained for time measurements with a hand-operated timing system.

(1.2) An electronic timer uses the mains frequency of 50 Hz as the basis of its timing. What type of accuracy would you expect for such a timer?

(1.3) Table 5.1 gives part of the specification for an electronic timer. Could such a timer be used for the following measurements?
(a) To measure a time of the order of 0.006 s to an accuracy of 0.0001 s.
(b) To distinguish between time intervals which might be as close as 1 ms in an overall time of about 5 s.
(c) To measure a time interval of about 200 ms to an accuracy of about ± 1%.
(d) To distinguish between time intervals such as 0.00110 s and 0.00105 s.

(1.4) Two oscillators are available with one marked as being stabilised and the other unmarked but assumed to be unstabilised. Which oscillator might be assumed to be the most accurate? Explain your answer.

ASSIGNMENTS

(1.1) Use a stopwatch to measure the time taken for ten complete oscillations of a simple pendulum of length 1 m. Repeat the measurement a number of times and determine the average time. What are the maximum deviations from this average time?

(1.2) Devise an electronic timing method for the measurement of the reaction time of a human to an event such as the coming on of a light. Measure reaction times and consider the spread of

results and to what factor or factors the spread can be attributed.
(1.3) Calibrate a clock or watch. Explain the standards used
and the method of calibration.

2. MEASUREMENT OF ANGULAR VELOCITY AND FREQUENCY

Angular velocity is defined as the ratio of the angular displacement to the elapsed time and has units of s^{-1}, or as sometimes written radians/s.

$$\text{Angular velocity} = \frac{\text{angle covered}}{\text{time taken}}$$

The angular velocity of, say, a rotating shaft is related to the frequency of rotation of that shaft. A frequency f means a time of $1/f$ for one rotation and hence an angle covered of 2π radians in that time.

Thus   angular velocity $\omega = \dfrac{2\pi}{1/f}$

$$\omega = 2\pi f$$

The angular velocity is directly proportional to the frequency of rotation. Thus angular velocity is sometimes expressed in terms of the frequency, e.g. as revolutions per minute (rev/min) or per second.

Instruments used for the measurement of the angular velocity of rotating shafts are called *tachometers*. There are two main types of tachometer—mechanical and electrical. One of the principal differences between the two types is power taken from the rotating shaft—mechanical types in general take more power from the rotating shaft. This means that the shaft speed of rotation might be significantly reduced when a mechanical tachometer is used.

The most common type of mechanical tachometer is that based on the *Watt governer* principle. The governer was a device invented by James Watt, about 1788, to control the speed at which the shaft of a steam engine rotated. *Figure 5.3* shows one of the forms used as a tachometer, a version for use on vertical shafts. The two masses A and B are attached to pivoted arms which rotate with the shaft. The arms

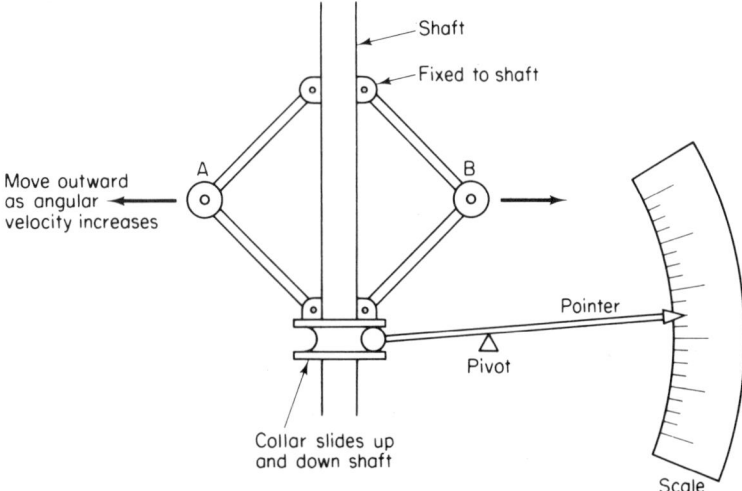

Figure 5.3 A mechanical tachometer

(a)   (b)

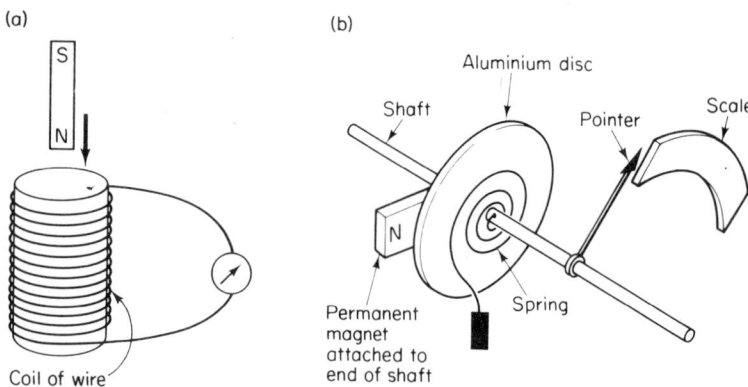

Figure 5.4 (a) Movement of the magnet relative to the coil, or the coil relative to the magnet, results in a current (b) Eddy current tachometer

Table 5.2 Typical example of the specification of a tachometer.

Based on speedometer principle with mechanical and magnetic coupling. For both left-hand and right-hand drive, reading 0–500 rev/min directly, 0–5000 rev/min (×10), 0–50 000 rev/min (×100), with range selector switch, stop and zero-setting buttons. Accuracy ± 0.5%. Temperature compensating device.

are acted on by a force, the weight of the collar, which tends to straighten them and keep the arms close to the shaft. When the shaft rotates the masses A and B move outwards, and the greater the angular velocity of the shaft the further out they move. This movement of A and B results in the collar sliding up the shaft which causes the pointer to move across the scale. The position of the pointer on the scale thus depends on the angular velocity of the shaft.

When there is relative motion between magnet and a conductor, currents are induced in the conductor. The effect is called electro-magnetic induction (*Figure 5.4a*). The *eddy current tachometer* depends for its action on this effect. A permanent magnet is attached to the rotating shaft (*Figure 5.4b*) and moves past a sheet of aluminium, the conductor. *Eddy currents* are induced in the aluminium. The currents flowing in the aluminium sheet produce their own magnetic field, like any current flowing in a conductor. The magnetic field is in such a direction as to cause the aluminium sheet to attempt to rotate and follow the moving magnet. The aluminium sheet however rotates against the torque produced by a spiral spring. The result is that the aluminium sheet rotates against the spring until the torque produced by the spring becomes opposite and equal to the torque produced by the interaction between the rotating magnet and the eddy currents in the aluminium. The size of the rotation depends on the angular velocity of the shaft, the eddy currents being greater the greater the angular velocity. The angle through which the aluminium sheet rotates can be indicated by a pointer moving against a scale.

The eddy current tachometer can be used for frequencies for direct measurements of rotating shafts up to about 200 revolutions per second with an accuracy of the order of a few per cent. A gear train can be introduced between the shaft and the tachometer for higher frequencies. This type of instrument is used in automobiles with a scale calibrated in linear velocity though the instrument is really measuring the angular velocity of rotation of the car wheels.

A *tachogenerator* used for the measurement of angular velocity is essentially a small electricity generator. The simple generator consists of a coil which rotates in a magnetic field (*Figure 5.5*). The faster the coil rotates the greater the e.m.f. induced in the coil. If electrical connections are made to the coil and the output fed to a voltmeter then the reading of the voltmeter is related to the angular velocity of the coil. Generators can have a coil rotating in a magnetic field, as in *Figure 5.5*, or have a magnet rotating past coils. Tachogenerators can be used for direct measurements of frequencies of rotating shafts up to about 100 revolutions per second with an accuracy generally of a few per cent.

There are a number of forms of *digital signal pick-up tachometer*. A transducer is used to give a digital output, the frequency of the digital output being related to the angular velocity. *Figure 5.6a* shows an *inductive pick-up tachometer*. A wheel with cogs is attached to the rotating shaft. As the cogs on the wheel move past the magnet a change in the magnetic flux through the coil is produced and this results in an e.m.f. being induced within the coil. Each cog moving past the magnet results in a voltage pulse in the coil. The frequency of these pulses, i.e. the number of pulses produced per second, is related to the angular velocity. *Figure 5.6b* shows one version of a *photoelectric pick-up tachometer*. A chopper disc is attached to the shaft and as the shaft rotates the light beam is 'chopped'. This results in the photoelectric

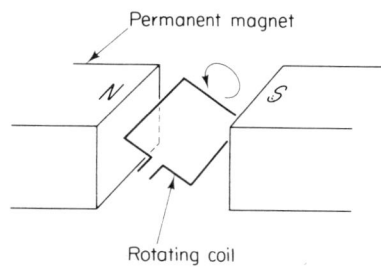

Figure 5.5 The basic elements of a generator

(a)

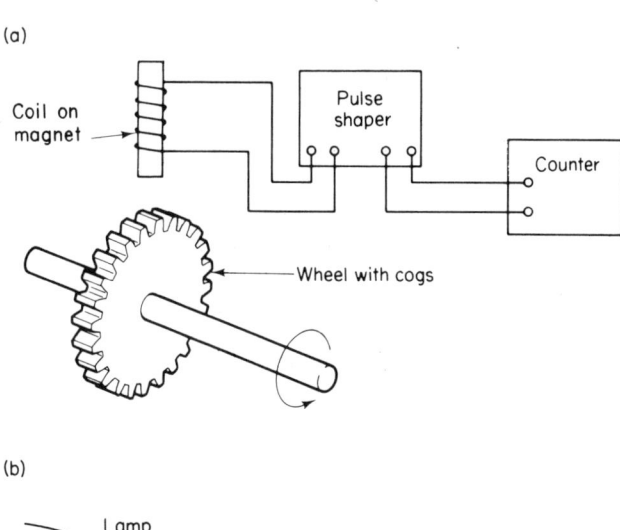

(b)

Figure 5.6 (a) An inductive pick-up tachometer (b) Photoelectric tachometer

cell detecting pulses of light, the frequency with which it receives pulses being related to the angular velocity of the shaft.

The pulses produced with either the inductive pick-up or the photo-electric pick-up generally have to be shaped so that they can be counted. The measuring system thus has a transducer connected via a signal conditioner, the pulse shaper, to a digital display. Such tachometers apply little or no load to the rotating shaft as there is no physical connection between the measuring system and the shaft. Frequencies as high as 100 000 revolutions per second can be measured using such a system.

An instrument which places no load on the rotating shaft is the *stroboscope*. Basically the stroboscope is a flashing light, the frequency of the flashes being controlled and measured. If the shaft has a mark made on it and the mark is viewed by means of the flashing light then if the frequency of the flashes is the same as the frequency of the shaft the mark will appear to be stationary because the shaft and the mark are only illuminated when the mark is in the same position on each revolution. When the mark appears to move it is due to the frequency of the stroboscope being different from the frequency of rotation of the shaft. Thus a measurement of the frequency of the flashing light gives a measure of the frequency of rotation of the shaft.

The mark on the shaft will appear to be stationary when the frequency of rotation of the shaft is a whole number multiple of the flash frequency. This is because the shaft may make any number of rotations between flashes and will appear to have the mark in the same place provided that in the interval between flashes the shaft makes a whole number of complete revolutions. Thus if a shaft is rotating with a frequency of 120 revolutions per second then it will appear stationary when the flashing light has a frequency of 120 Hz, 60 Hz or 40 Hz, etc. At 120 Hz the shaft makes one revolution in the time between flashes. At 60 Hz the shaft makes two revolutions between flashes. At 40 Hz the shaft makes three revolutions between flashes.

If the frequency of rotation of the shaft is some simple submultiple of the flash frequency 'multiple images' of the single mark on the shaft will be seen to be apparently stationary. Thus if the shaft rotates with a frequency of 120 revolutions per second it will appear to have a single stationary mark when the flashing light has a frequency of 120 Hz, two stationary marks when the frequency is 240 Hz, three stationary marks when the frequency is 360 Hz. At 240 Hz the mark is at the same position only every second flash, being at an intermediate position at the intermediate flash. At 360 Hz the mark is at the same position only every third flash, being at intermediate positions at the intermediate two flashes.

Thus for a single mark on the shaft to seem to be stationary:

$$\text{shaft frequency} = \text{flash frequency}$$

or

$$\text{shaft frequency} = k \times \text{flash frequency}$$

where

$$k = 1, 2, 3, 4, \text{etc.}$$

or

$$\text{shaft frequency} = \frac{\text{flash frequency}}{k}$$

where $k = 1, 2, 3, 4$, etc. In this last case multiple images are produced. To find the frequency of rotation of a shaft a possible method is to gradually increase the stroboscope frequency until the highest frequency is found at which a single image of the single mark is seen. This frequency is the shaft frequency.

Commercial stroboscopes usually have a neon or xenon tube as the light source and operate at frequencies in the approximate range 10 to 300 Hz.

The advantages and disadvantages of the different methods for the measurement of angular velocity discussed in this section can be summarised as follows.

Watt governer systems: imposes a load on the shaft, rough measurement, can be used to control angular velocity (see chapter on control), robust.

Eddy current tachometer: Very little load imposed on the shaft, range normally up to about 200 revolutions per second, gives a display at a distance, relatively cheap.

Tachogenerator: Very little load imposed on the shaft, range up to about 100 revolutions per second, gives a display at a distance, similar accuracy to eddy current tachometer, relatively cheap.

Digital signal pick-up tachometers: Very little load imposed on the shaft, ranges up to about 100 000 revolutions per second, gives a display at a distance, can be accurate, not cheap.

Stroboscope: No load imposed on the shaft, generally range up to 300 revolutions per second, the shaft has to be observed, reasonably accurate.

Table 5.3 Typical example of a manufacturer's specification for a xenon stroboscope.

Flashing rate:	Range 1 60 to 600 flashes per minute
	Range 2 600 to 6000 flashes per minute
	Range 3 3000 to 15000 flashes per minute
Mean flash power:	6 watts approximately.
Accuracy:	± 3% full scale deflection on each range.
Flash duration:	5 to 10 microseconds
Triggering:	Internal oscillator
	External shorting contacts
Power supply:	210 to 240 V a.c. 50 Hz
	110 V version available to order.
Dimensions:	225 × 175 × 175 mm
Weight:	3 kg

PROBLEMS

(2.1) *Table 5.2* gives the specification for a tachometer.
(a) Could the instrument be used to measure a change in rate of rotation of a shaft from 200 rev/min to 202 rev/min?
(b) What is the maximum rate of revolution of a shaft that can be measured?
(2.2) Which type of system could be used for the measurement

of the angular velocity of a shaft if no load had to be placed on the shaft?

(2.3) Explain how a counter could be used with an appropriate transducer to give a tachometer.

(2.4) A tachometer is available which can measure frequencies of rotation of a shaft up to 100 revolutions per second. What gear train could be used between the shaft and the tachometer to enable it to measure up to 1000 revolutions per second?

(2.5) Explain how you would use a stroboscope to measure the frequency of rotation of a shaft.

(2.6) *Table 5.3* gives the specification for a stroboscope.
(a) On range 1 by how many flashes per minute could the indicated frequency be in error?
(b) Could the stroboscope be used to measure a frequency of 100 Hz?

ASSIGNMENTS

(2.1) For a given rotating shaft specify the most appropriate measurement system to enable the frequency of rotation to be determined, measure the frequency and then comment on the accuracy of the result.

(2.2) Devise a photoelectric pick-up tachometer suitable for the measurement of frequencies of rotation of the order of 1000 rev/min.

3. COMPARATORS   *Comparators* are instruments which enable a comparison to be made between the item being measured and a length standard. They can also be used to detect changes in length. All comparators tend to consist of three basic items:

(1) Some device, generally a plunger, which senses the change in length or difference in length.

(2) A signal conditioner which magnifies the small displacements of the plunger.

(3) A display system.

The range of a comparator is generally quite small, depending on the magnification of the system, the greater the magnification the smaller is the range. A comparator with a magnification of the order of 500 might have a range of about 0.3 mm; a similar comparator with a magnification of 3000 might have a range of only 0.05 mm. Comparators are generally sensitive to changes of the order of 0.002 mm or less.

An important factor affecting the accuracy of a comparator is the pressure exerted by the plunger on the item being compared. As far as possible this measuring pressure should be the same for all readings. If this is not the case the instrument reading may vary when repeated measurements are made of the same object. In addition the pressure applied should be low. If this is not the case some deformation of the object being measured may occur.

There are many forms of comparator: mechanical, pneumatic, optical and electric. *Figure 3.8*, the *Sigma comparator*, and *Figure 3.25*, the *Johansson comparator*, are examples of mechanical comparators. The first employs a compound lever system and the second a twisted strip as signal conditioner to give a large magnification of the movement of the plunger. *Figure 2.8* gives the basis of a pneumatic

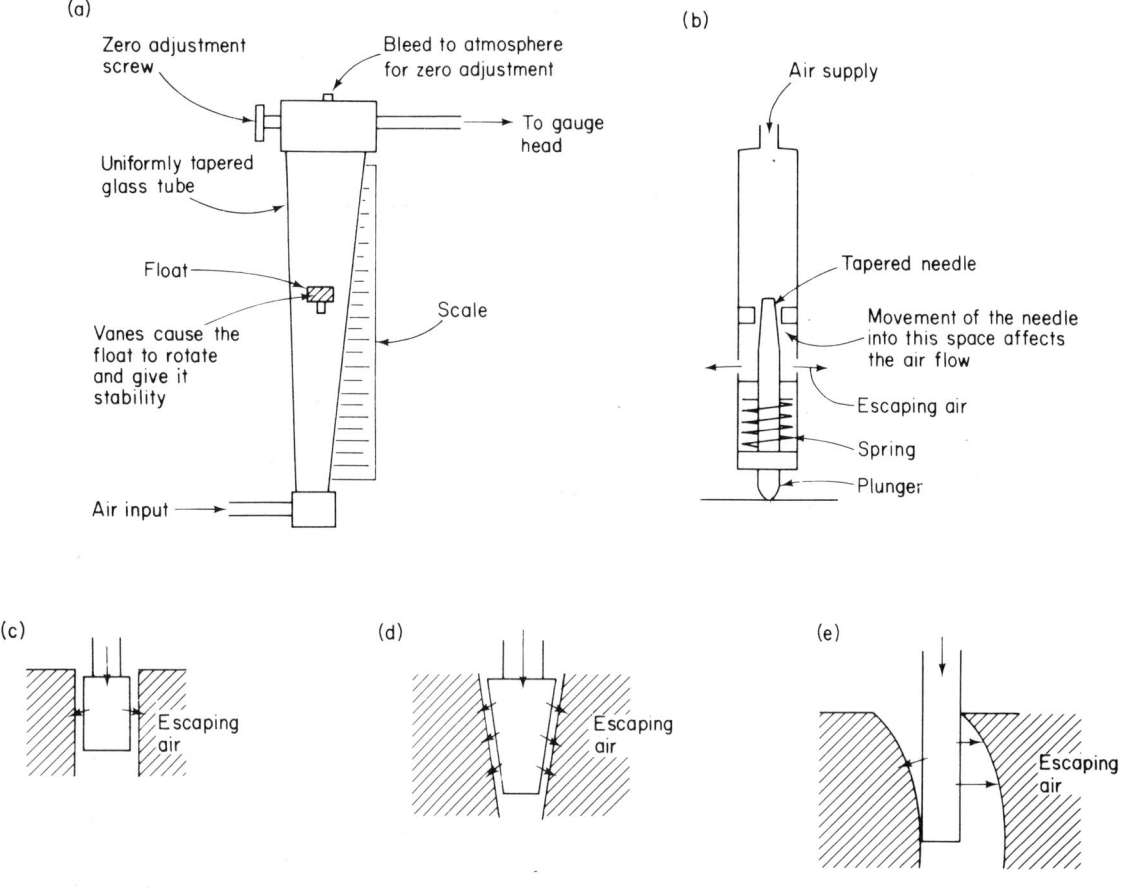

Figure 5.7
(a) A pneumatic comparator. The position of the float depends on the velocity of the air flow
(b) One form of gauge head for comparison of displacement
(c) Form of gauge head for measurement of hole diameter
(d) Form of gauge head for measurement of hole taper
(e) Form of gauge head for measurement of bore straightness

comparator. *Pneumatic comparators* work on the principle that if an air jet is in close proximity with a surface the flow of air out of that jet is restricted. This can result in a change of pressure in the system and a change in the rate of flow of air through the system supplying the jet. The instrument illustrated in *Figure 2.8* depends on a measurement of the change in pressure. *Figure 5.7* shows a pneumatic comparator based on a measurement of the change in the rate of flow. *Figure 2.9* shows two examples of *electrical comparators*.

Comparators are used to make comparisons between the object being measured and a standard. The standards normally used are *end standards*. With an end standard the length is defined for the distance between the two opposite faces or ends. Such standards are called *gauge blocks* and *length bars*. Gauge blocks are used for lengths up to 200 mm and length bars for greater lengths.

Gauge blocks are rectangular hardened high carbon steel blocks with the two opposite faces, the defined distance apart, being ground and lapped to make them flat and parallel within defined limits. The flatness and finish of the surfaces are such that the blocks can be made to adhere to each other in order to build up a stack of blocks to give a wide variety of different length standards from a relatively small number of blocks. The process of making blocks adhere to each other is known as *wringing*.

Gauge blocks are available in five grades:

Grade 00: These are supplied with a calibration chart and are used as the standards by gauge block manufacturers. The grade of accuracy of these blocks is very high and is not normally necessary for normal routine requirements.

Calibration grade: This grade of block is intended for the calibration of other blocks in normal engineering practice and is not intended for general gauge inspection. A calibration chart is provided.

Grade 0: This grade is intended for inspection and work of high precision.

Grade I: This is the general purpose gauge grade and is used for measurement in production.

Grade II: This grade is used for rough purposes, being the least accurate grade of gauge block.

To illustrate the different tolerances, i.e. accuracies, of gauge blocks of different grade, the following are the length tolerances for gauge blocks between 80 and 100 mm long.

Grade 00	± 0.15 μm and calibration chart supplied
Calibration grade	± 0.50 μm and calibration chart supplied
Grade 0	± 0.25 μm
Grade I	$+ 0.60$ μm to -0.30 μm
Grade II	$+ 1.40$ μm to -1.00 μm

$(1 \ \mu m = 10^{-6} \ m = 10^{-3} \ mm)$

The Grade 0 has a closer tolerance than the Calibration grade, the Calibration grade has however a calibration chart. Grade I and II do not have tolerances equally disposed about the nominal size as they are expected to suffer wear with use, while the other grades are not expected to suffer so much wear.

Length bars are steel cylinders approximately 22 mm in diameter and are available in a number of lengths. The bars are available in a number of different grades.

Reference grade: Only used where the very highest accuracy is required. A calibration chart is supplied.

Calibration grade: This grade is intended for calibration. A calibration chart is supplied.

Inspection grade: These are intended for use in inspection.

Workshop grade: For general use in workshops.

The different tolerances for the various grades of length bar are as follows:

Reference grade	± 0.08 μm and calibration chart supplied.
Calibration grade	± 0.15 μm and calibration chart supplied.
Inspection grade	± 0.18 μm
Workshop grade	± 0.30 μm

Gauge blocks and length bars are quoted as being at their true length at a temperature of 20 ± 0.5 °C. Thus for accurate measurements the temperature at which the gauge blocks or length bars are used needs to be controlled.

The advantages and disadvantages of the various forms of comparator discussed in this section can be summarised as follows.

Mechanical comparators: magnifications up to about × 5000 are generally possible, permitting measurements with the most sensitive instruments to about 0.0001 mm.

Pneumatic comparators: a supply of clean compressed air is required, no mechanical contact is made with the component being measured, can be adapted for internal diameters, magnifications up to about × 40 000 are generally possible.

Electrical comparators: a power supply is required, a low measurement pressure is applied to the object being measured, fast response to changes, magnifications up to about × 50 000 are generally possible.

PROBLEMS

(3.1) Explain the principle of operation of a pneumatic comparator.

(3.2) Explain the principle of operation of an electrical comparator.

(3.3) Which type of comparator would you suggest for the measurement of the internal diameter of a hole?

ASSIGNMENTS

(3.1) Make a critical analysis of the requirements of some aspect of industry for measurements using a comparator and specify the most suitable comparator or comparators for the work in question.

4. MEASUREMENT OF LENGTH

For the type of measurement of length where the quantities being measured vary quite considerably then the *micrometer screw gauge* or *vernier calipers* will most likely be the best form of measuring system if a small measurement with reasonable accuracy is required. If the measurement of a particular quantity needs repeatedly to be made then a comparator will probably be the most convenient form.

The basis of the micrometer is to produce linear movement of an anvil, which presses against the item being measured, by means of the

Table 5.4 Typical example of the specification of a micrometer screw gauge.

Range 0–25 rnm × 0.01 mm, precision made, with hardened and ground screw, enamelled I-section frame; with ratchet and lock nut.

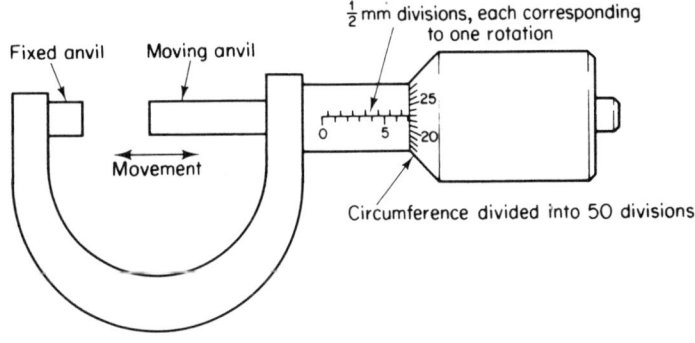

Figure 5.8 Micrometer screw gauge

rotary movement of a screw thread. The amount of rotary movement is much greater than the linear movement. Thus if the thread has a $\frac{1}{2}$ mm pitch then one complete rotation through 360° of the screw results in a linear movement of $\frac{1}{2}$ mm. If the circumference of the screw head is divided into 50 divisions a movement which involves the head rotating by one division means a linear motion of 1/50th of 0.5 mm, i.e. 0.01 mm. *Figure 5.8* shows a typical micrometer screw head. The accuracy of the typical micrometer is of the order of ± 0.01 mm. A limiting factor on the accuracy is the difficulty in obtaining the same pressure of the anvil on the specimen being measured every time a reading is taken.

The magnification of the vernier type of scale is achieved by having two scales sliding over each other, one of the scales having divisions smaller than the other. The divisions that are in line are then detected. *Figure 5.9* shows a typical vernier scale. The smallest unit of length that

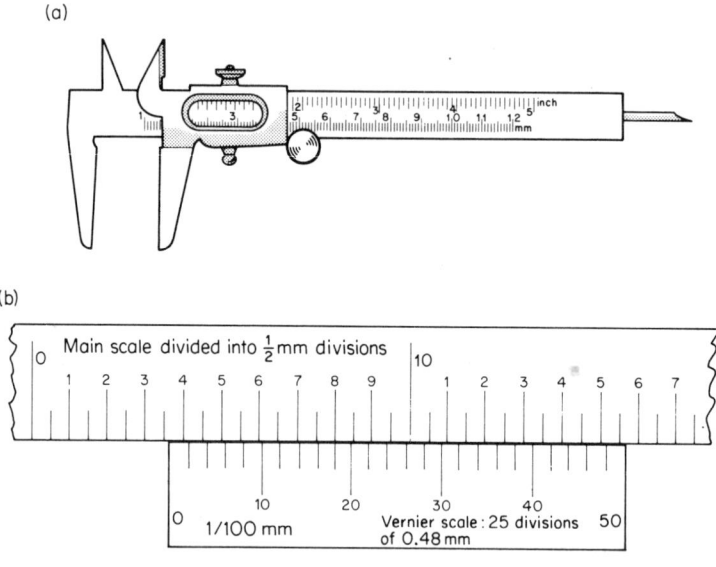

Figure 5.9 (a) Vernier calipers (b) A vernier scale

Table 5.5 Typical example of the specification of vernier calipers.

Nickel plated steel, double scale, 0–12 cm × 0.1 mm; bevelled cross horns for internal measurements; sliding jaw, manipulated by fine adjustment thumbwheel.

can be read with a vernier scale is equal to the difference in length between the divisions on each scale. Thus for a scale to read to 0.02 mm then the divisions on the vernier scale must be 0.02 mm smaller than the divisions on the main scale. Thus with main scale divisions of 0.50 mm the divisions on the vernier scale must be 0.50−0.02 = 0.48 mm. Since the 0.50 mm division of the main scale has to be divided into 0.02 mm units by the vernier scale there must be 25 divisions on the vernier scale.

If the zero of the vernier scale is in line with a mark on the main scale then the reading is that due to that mark on the main scale. If the vernier scale is then displaced so that its first mark lines up with the next mark up the main scale then the displacement has been 0.02 mm. If the vernier scale is displaced so that its second mark is in line with a division on the main scale then the displacement has been 2 × 0.02 mm. If the seventh mark on the vernier scale lines up with a mark on the main scale then the displacement has been 7 × 0.02 mm.

The reading is obtained as follows:

(1) Read the main mm units on the main scale immediately prior to the vernier zero mark.

(2) Add to that reading any 0.5 mm unit that occurs prior to the vernier zero mark.

(3) Add to that total the vernier reading indicated by the mark on the vernier scale which is exactly opposite a division on the main scale.

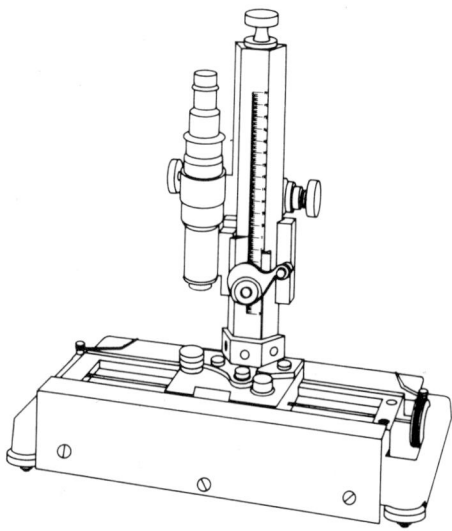

Figure 5.10 Vernier microscope (reproduced by permission of Griffin & George Ltd)

Table 5.6 Typical example of a specification of a vernier microscope similar to that shown in *Figure 5.10*.

Scale length 16 cm, reading to 0.01 mm.
Vertical and horizontal scales reading to 16 cm, with verniers; heavy cast base, synthetic (hammer) finish with vee-shaped slide, scale graduated in mm; vertical column mounted on sliding carriage, with rack and pinion focusing microscope and vernier scale; Ramsden eyepiece 22 mm focal length, equivalent magnification × 12.5; circular scale on microscope mount graduated 0–90° enabling repeatable setting of the microscope to be made between vertical and horizontal; spring-loaded ball catches with locking clamp to locate microscope in the vertical and horizontal positions; both sliding carriages are clamped to push-rods actuated by fine pitch screws for fine adjustment; mounted magnifier to read vertical scale, hand magnifier for horizontal scale; levelling feet and spirit level; complete with Ramsden eyepiece with cross-wires and objective in polished wood case.

The instrument is standardised at 20°C and the coefficient of linear expansion of the material of the scale is 0.000 017 per deg C over the range 25–300°C.

Available with objectives 25 mm, 50 mm, 75 mm.

The range over which micrometer screw gauges are typically used is about 0 to 25 mm, vernier calipers can be used up to about 200 mm. For the determination of lengths greater than this, travelling microscopes can be used. In such instruments the position of each end of the length being measured is compared against a scale by the use of a microscope, the scale generally being a vernier scale. *Figure 5.10* shows a typical vernier microscope instrument.

PROBLEMS

(4.1) For what type of length measurements are (a) comparators, (b) micrometer screw gauges, (c) vernier calipers, (d) travelling microscopes, likely to be the optimum choice?

(4.2) The micrometer screw gauge referred to in the specification in *Table 5.4* has a ratchet and lock nut. What is the purpose of these items?

(4.3) What is the percentage of full scale accuracy possible with the vernier calipers specified in *Table 5.5*?

(4.4) Describe how vernier calipers could be used to measure the internal diameter of the end of a pipe.

(4.5) Explain the statement, in the specification of the vernier microscope in *Table 5.6*, of the temperature at which the instrument is standardised and the significance of the quoted coefficient of linear expansion.

(4.6) Explain how you would check the calibration of a micrometer screw gauge.

ASSIGNMENTS

(4.1) Check the calibration of either a micrometer screw gauge or vernier calipers and produce a calibration chart for the instrument.

(4.2) Devise a vernier scale for use with a steel rule, marked in millimetres, which would enable you to read a displacement to 0.1 mm.

5. DISPLACEMENT AND LEVEL MEASUREMENT SYSTEMS

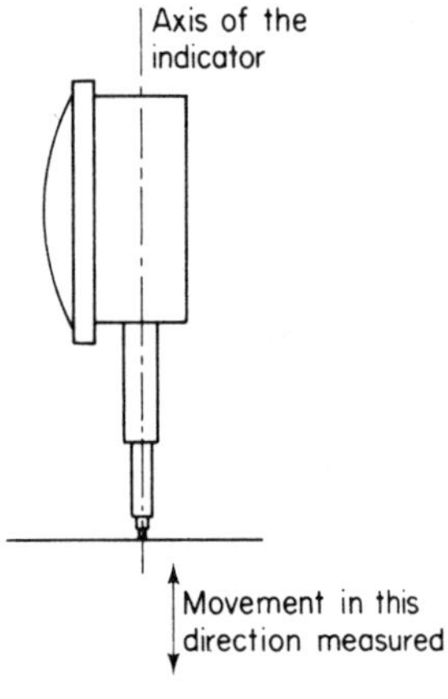

Figure 5.11 Dial test indicator

A *dial test indicator* arrangement (*Figure 5.11*) offers a method of determining the displacement of an item. Movement of the item results in a movement of the plunger of the dial test indicator and hence a reading on that instrument. Such a system enables a continuous moni-

Table 5.7 Specification of a typical dial test indicator.

Continuous metric calibration. Precision movement in rigid alloy case with precision-ground stainless steel stem and rack pinion; spindle fitted with ball anvil; white dials, figured black, adjustable through 360° and fitted with locking clamp; revolution counter dial; back fitted with back lug offset from centre line, with fixing hole.
Range 12 mm, subdivisions 0.01 mm, graduations 0–100.

toring of the displacement to be made.

The displacement must however take place along the axis of the indicator if it is to record the true displacement.

Comparators are instruments that can be used to monitor displacements. Both they and the dial test indicator are however concerned

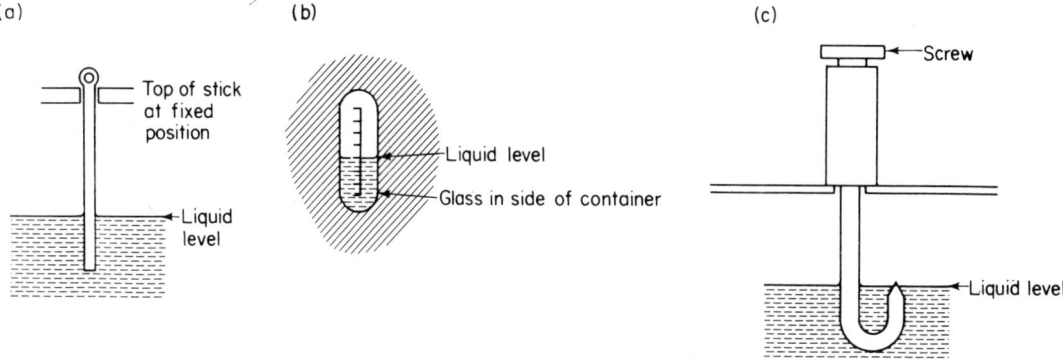

Figure 5.12 Forms of dip-stick (a) Dip-stick (b) Gauge glass (c) A hook gauge

with the displacement of solid surfaces. Different systems are used for the determination of the displacement of liquid surfaces, i.e. level measurement.

A *dip-stick* is a simple method of determining liquid level (*Figure 5.12a*). The height up the stick at which the liquid surface occurs is shown by a deposit of the liquid on the stick when it is removed from the liquid, the stick top always being at the same position when it is in the liquid. This method is used to check the oil level in the motor car engine sump. Dip-sticks have the disadvantage of needing to be removed and examined for the level of the liquid to be determined.

One form of dip-stick which overcomes this disadvantage is the use of a *gauge glass* in the side of the container. The level of the liquid surface can be noted against a scale on the glass (*Figure 5.12b*). This is not, however, a suitable method if the liquid leaves a permanent opaque film on the glass and so renders it impossible to see the liquid level.

The *hook gauge* (*Figure 5.12c*) is another form of dip-stick. The height of the hook is adjusted by means of a screw until the point on the end of the stick is just in contact with the liquid surface. The movement of the screw controls the movement of a marker across a scale.

There are a number of methods based on the use of buoyant floats (*Figure 5.13*). The float is often a metal hollow sphere attached to the

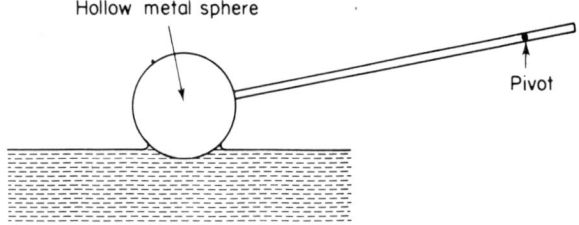

Figure 5.13 Buoyant float

end of a lever system. Changes in the level of the liquid cause the float to move and hence cause movement of the lever. The lever could be used itself as the indicating mechanism or could be used to provide an input to a signal conditioner.

One method of level determination which does not require any mechanical contact with the surface is the *ultrasonic level indicator*

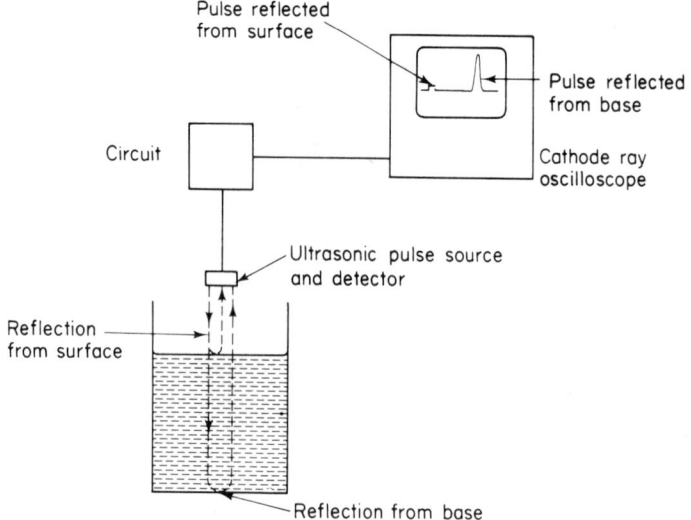

Figure 5.14   Ultrasonic level indicator

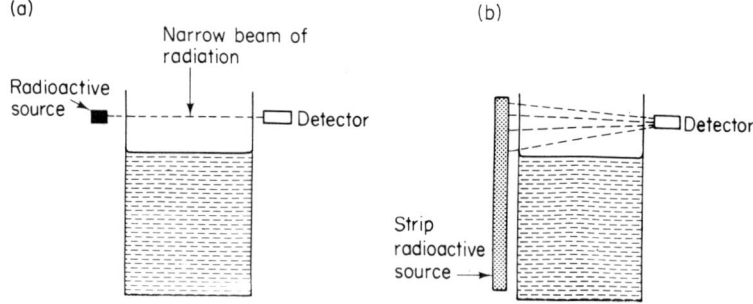

Figure 5.15   (a) On-off level radioactive gauge (b) Level indicator radioactive gauge

(*Figure 5.14*). This is essentially an echo sounder. A pulse of ultrasonic waves is emitted at a source above the liquid surface. The pulse travels down to the liquid surface where part of the pulse is reflected. The remainder of the pulse continues down through the liquid before being reflected from the base of the container. Both these reflected pulses are detected and the difference in time between the reception of the pulses is a measure of the depth of liquid in the container. The pulses are often displayed on a cathode ray oscilloscope. This method can be used with powders as well as liquids.

   Radioactive isotopes can be used for level measurement systems where no mechanical contact is made with the liquid surface. *Figure 5.15a* shows a simple on/off level determination system. The intensity of the radiation, generaly gamma radiation, from the source is significantly reduced when the liquid comes between the source and the detector. The output from the detector can thus be used to give an indication of whether the liquid is between the source and detector or not. *Figure 5.15b* shows a system which can be used to give an output related to the position of the liquid level. Instead of using a small radioactive source producing a narrow beam of radiation, as in the on/off detector, a strip source is used. When the liquid level changes the amount of radiation reaching the detector changes, the output from the detector thus becomes a measure of the position of the liquid level.

There are many methods used for the determination of level in a container which depend on a measurement of the *weight* of the container, the more liquid in the container the greater the height of the surface above the base and the greater the weight. Such methods can be used for both liquids and solids. The section in this chapter on the measurement of force gives details of load cells, a common method used for measurement of the weight of containers.

With liquids *pressure measurements* can be used as a measure of the height of liquid. The pressure exerted at the base of a column of liquid of height *h* and density *d* is *hdg*, where *g* is the acceleration due to gravity. *Figure 5.16* shows a system based on a measurement of the difference in pressure between the lower and upper parts of the container.

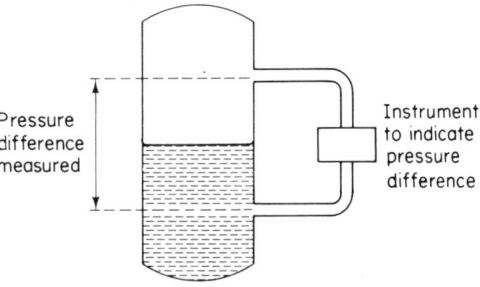

Figure 5.16 Differential pressure measurement for liquid level determination

The advantages and disadvantages of the various forms of liquid level measurement systems discussed in this section can be summarised as follows.

Dip-stick: Very rough measurement, has to be examined for the measurement to be made, cheap.

Gauge glass: Rough measurement, not suitable with liquids that leave a film on glass, cheap.

Hook gauge: More accurate than the dip-stick, has to be adjusted before a reading can be made.

Float systems: Can give a display, not accurate, cheap, can be used to control the level (see the Chapter on control).

Ultrasonic level indicator: Gives a display at a distance, can be used with liquids and powders, can be used over quite wide ranges of depth, reasonably accurate, not cheap.

Radioactive isotope methods: Give a display at a distance, can be used with liquids and powders, often only on/off indicators, not cheap.

Weight systems: Can give a display at a distance, can be used with liquids and powders (see the section on force measurement), can be used with very large containers, can be accurate.

Pressure measurement systems: Can be used with liquids under pressure in containers.

PROBLEMS

(5.1) How would you check the calibration of a dial test indicator?

(5.2) What is the percentage of the full scale reading accuracy for the dial test indicator specified in *Table 5.7*?

(5.3) Under what circumstances would a buoyant float system be preferable to a dip-stick?

(5.4) Explain the principle of the ultrasonic level indicator.

(5.5) What type of measuring system might be suitable for the determination of the level of liquid in a fire extinguisher? Give reasons for your answer.

(5.6) What type of measuring system might be suitable for the determination of the level of the water in a cold water tank mounted in the roof of a factory? Give reasons for your answer.

ASSIGNMENTS

(5.1) Devise a float system which can be used to give a display related to the water level in a container. Test and calibrate the system. The level is required to an accuracy of 5 mm, the range of level variation being of the order of 200 mm.

(5.2) Devise and calibrate a differential pressure system for the measurement of liquid level in a container.

6. MEASUREMENT OF FORCE

The basic unit of force is the *newton* (N), this being the force which would give a mass of 1 kg an acceleration of 1 m/s². *Gravitational forces* are exerted on masses due to the interaction between the object concerned and the earth (or indeed any other large mass). Close to the earth's surface the gravitational force exerted on a mass M is Mg, where

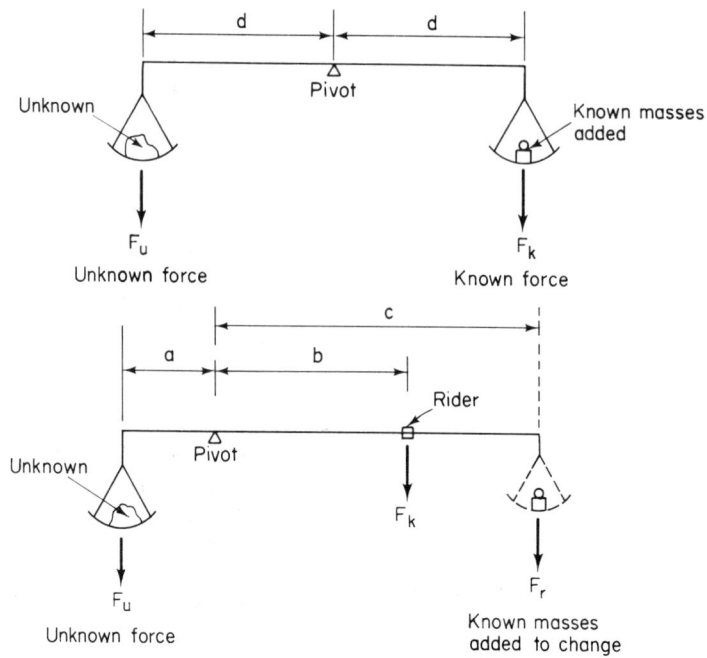

Figure 5.17 (a) A simple analytical balance (b) A simple mechanical balance

g has the value of the acceleration due to gravity (or gravitational field strength). This has a value of approximately 9.81 m/s², the value depending however on the locality. Thus a mass of 1 kg has, close to the earth's surface, a gravitational force acting on it of about 9.81 N. The *weight* of an object is the force that has to be applied to cancel the gravitational force. If there were only the gravitational force acting on an object then it would fall and accelerate, when it is stationary there is no net force acting on it. An object stationary on the pan of a spring balance has no net force acting on it and so the gravitational force has been cancelled by the pull of the spring, the pull of the spring thus being a measure of the weight of the object. An object of mass 1 kg stationary at the earth's surface has a weight of about 9.81 N. *Scales* and *balances* are examples of force measuring systems. *Figure 5.17a* shows the form of the simple analytical balance that might be used for weighing chemicals. The gravitational force acting on an unknown mass is compared with that acting on a standard mass. The balance is a lever with the known and unknown forces being equidistant from the pivot. As the moment of the known force about the pivot

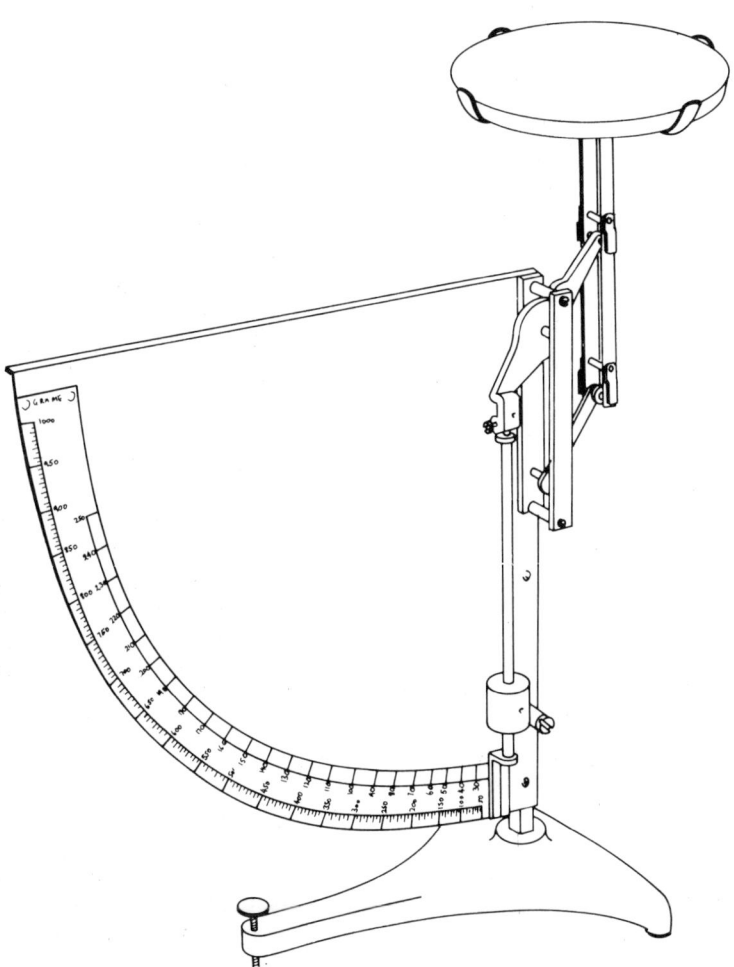

Figure 5.18 Lever arm balance

must equal the moment of the unknown force about the pivot and they are both the same distance from the pivot then the known and unknown forces must be equal at the balance condition when the balance is in equilibrium.

Moment of known force about pivot $\quad = F_k \times d$

Moment of unknown force about pivot $= F_u \times d$

At balance the two moments are equal, hence

$$F_k \times d = F_u \times d$$

$$F_k = F_u$$

The balance condition is obtained by adding masses at one end of the lever until F_k becomes equal to F_u.

Figure 5.17b shows another form of balance. In this balance equilibrium is achieved by sliding a mass along the lever arm. The range of the instrument can be changed by adding masses to the end of the lever arm. This is done to obtain a rough balance and then the fine balance is achieved by sliding the mass along the lever arm.

At equilibrium

$$F_u \times a = F_r \times c + F_k \times b$$

The position of the rider along the lever arm, i.e. distance b, can therefore be used as a measure of force.

Figure 5.18 shows another form of balance depending on lever action. The unknown mass is placed on the balance pan and the gravitational force acting on it causes the lever arm to swing out over the scale. The scale is calibrated in terms of mass. The balance is known as the *lever arm balance* and generally has two ranges. The ranges can be selected by changing the position of the rider on the lever arm.

The balances shown in *Figures 5.17, 5.18* and *5.19* all depend on the lever. Another group of balances, and force measuring systems, depend on the stretching of some elastic member. A *spring balance* depends for its action on the stretching of a spring under the action of force, the extension of the spring being proportional to the force (*Figure 5.19*). Direct reading spring balances are not capable of high accuracy as the extensions produced are relatively small.

Steel rings (*Figure 5.20*), or *proving rings* as they are generally called, can be used for the accurate measurement of force. A force applied across a diameter causes a distortion of the ring, the amount of distortion being proportional to the force. For relatively rough measurements the distortion can be measured with a dial test indicator gauge, for more accurate methods micrometer screws or displacement transducer systems can be used. The distortion produced as a result of the force is relatively small and the accuracy of the system depends very much on the accuracy with which the distortion can be measured.

Hydraulic pressure transducers may be used for the measurement of force. *Figure 5.21* shows one form of such a system. A chamber containing oil is connected to a pressure gauge, possibly a Bourdon tube gauge. The chamber has a diaphragm to which the force is applied. The applied force causes a change in pressure of the oil which then shows up as a change in reading on the pressure gauge. The gauge can be calibrated directly in terms of the applied force.

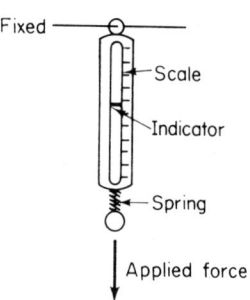

Figure 5.19 Spring balance

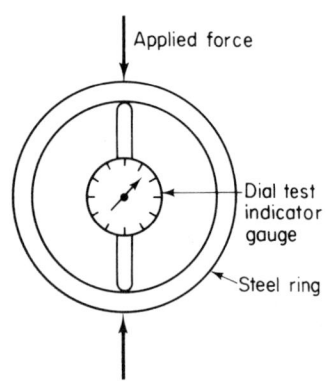

Figure 5.20 Proving ring

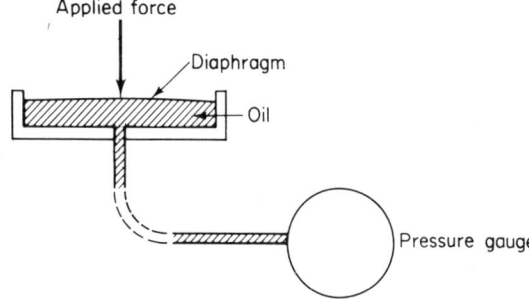

Figure 5.21 Hydraulic pressure force measurement system

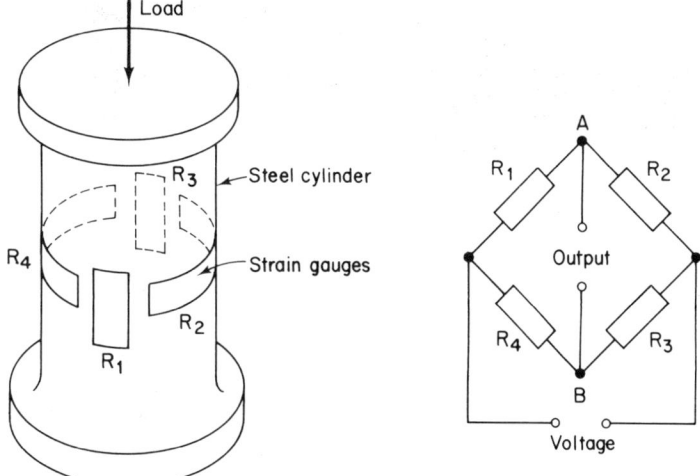

Figure 5.22 A load cell

Strain gauges may be used to determine the deformation of some elastic member when subject to force and so provide a measurement of the force. Such a system is known as a *load cell. Figure 5.22* shows one form of load cell. The elastic member to which the strain gauges are attached is, in this case, a steel cylinder. Four strain gauges are used, each gauge forming one arm of a Wheatstone bridge. When a compressive load is applied to the cylinder the strain gauges R_1 and R_3 suffer compression and so decrease in resistance while the strain gauges R_2 and R_4 suffer tension and so increase in resistance. When there is no load all the four gauges are the same resistance and so the potential difference between points A and B is zero. When subject to load an out-of-balance potential difference is produced which is related to the applied load. The use of four identical strain gauges, one in each arm of the bridge, eliminates the effect of temperature. A change in temperature affects equally each arm of the bridge and so produces no out-of-balance potential difference.

Figure 5.23a shows a load cell being used for the measurement of material level in a bin, the greater the height of material in the bin the greater the volume of the material and hence the greater the weight. Three load cells are used to support the bin and thus the force indicated by the cells is a measure of the level of the material in the bin.

Figure 5.23 (a) Typical horizontal tank installation. A 30 tonne chlorine tank supported on three load cells and stabilised horizontally by means of tie bars (b) Hydraulic jack with built in load cell (reproduced by permission of Davy Instruments Ltd)

Figure 5.23b shows a hydraulic jack with a load cell built into it. Three or four of these can be used to jack up a component and so determine the weight of the component, e.g. a giant transformer having a mass of say 300 tonnes.

Load cells have many applications. They are used in weighbridges for weighing lorries, in the hook of a crane to measure the load carried by the crane hook, for measuring the weight of a material stored in a silo or tank. Load cells are relatively compact and, because they are electrical, can give a display some distance away from where the cell is located. They are robust in that there are no moving parts. They can be used to measure loads from of the order of 50 kg to more than 5 000 000 kg. *Figure 1.3* shows some of the load cells commercially available.

Strain gauges can be used to give measuring systems which can be used for transient force or pressure measurements. *Figure 5.24* shows one form of such a system. Two gauges are attached to a diaphragm which deforms under the action of a force. The resistance of the gauges

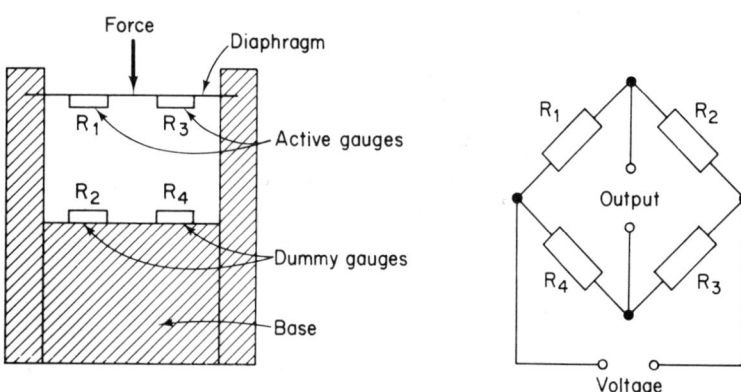

Figure 5.24 Force/pressure measuring system using semiconductor strain gauges

changes as a result of the deformation. Two other, identical, gauges are attached to a base plate which does not deform. All four gauges are in a Wheatstone bridge, one in each arm. The two gauges attached to the base respond only to temperature changes and thus cancel any temperature changes affecting the gauges on the diaphragm. The out-of-balance potential difference of the bridge is determined by the changes in resistance of the gauges and hence is related to the applied force.

The advantages and disadvantages of the force measurement systems discussed in this section can be summarised as follows.

Scales and balances: Can be direct reading or require balancing, used mainly for weighing, can be made very sensitive, can be made robust, high accuracy is possible, the ranges depend on the form of the instrument concerned, with the lower limit being the measurement of masses of the order of 0.001 g and the upper limit of the order of 1000 kg, the less accurate scales are relatively cheap.

Proving rings: Direct reading or can give a display at a distance, can be very accurate, range of the order of 2 kN to 2000 kN.

Hydraulic pressure transducer systems: Give a display at a distance, no power supply necessary, relatively rapid response to force changes, pressure differences of the order of 1000 Pa to 200 000 Pa can be measured depending on the form of the system (see section on pressure measuring systems).

Load cells: Give a display at a distance, power supply required, rapid response, robust, the ranges depend on the form of the cell with the lower limit being of the order of 500 N and the upper limit about 6000 kN, can have reasonable accuracy.

ELECTRONIC SCALE

with load cells in connector bolts

- Eliminates time-consuming transportation to permanent weighing stations
- Ensure accurate checking of material flow.
- Facilitate inventory-taking.
- Totalizing unit (checks day's production for example).
- Easy to install in most industrial trucks.

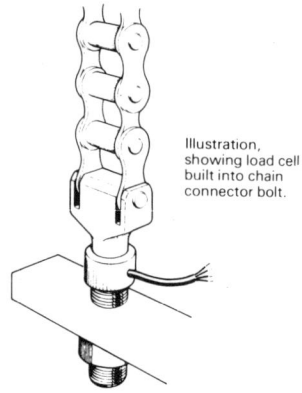

Illustration, showing load cell built into chain connector bolt.

Ergotest weighing system is designed for rugged material handling applications. Weights are obtained by recording tensile force in the lifting chain connector bolts. Weights are presented in digital form on a liquid crystal display that is easy to read outdoors. Designed to simplify both installation and adjustment (estimated to take 3 - 4 hours). Readout instrument can accommodate an optional totalizing and/or printing unit.

System can be preset (tared) to compensate for weight of packing materials etc. Result: net weight will be read directly on the instrument display.

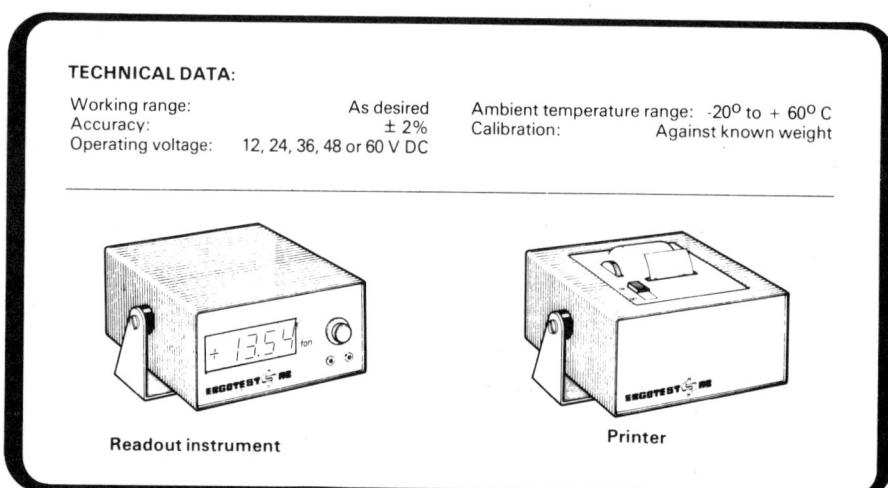

TECHNICAL DATA:

Working range:	As desired
Accuracy:	± 2%
Operating voltage:	12, 24, 36, 48 or 60 V DC

Ambient temperature range: -20° to + 60° C
Calibration: Against known weight

Readout instrument

Printer

C. Stevens & Son (Weighing Machines) Ltd.

LONDON & HEAD OFFICE

287/289 Goswell Road, London, EC1V 7LD (Regd Office)
Telephone: 01-837 8173/4/5. Telex: 22106 SCALES

ERGOTEST AB   BOX 382   S 851 06 SUNDSVALL   SWEDEN   TEL 060 12 16 25

Figure 5.25   (reproduced by permission of C. Stevens & Son Ltd)

PROBLEMS

(6.1) Explain the way in which the lever arm balance operates.

(6.2) Which type or types of measuring system would you consider appropriate for incorporation in a tensile test machine for the measurement of the force exerted on standard metal test specimens? Give reasons for your answer and specify the range of force you would expect to be measuring and the accuracy.

(6.3) Explain how load cells can be used as a basis of a measuring system for the level of liquid in a large container, as in *Figure 5.23a*.

(6.4) *Figure 5.25* shows a page from a manufacturer's catalogue in which a load measuring system is described. Explain the principles of the measuring system and the way in which it is intended to be used.

(6.5) Load cells can use electrical resistance strain gauges. How can the effects of temperature be compensated for with the strain gauges when used with load cells?

ASSIGNMENTS

(6.1) A simple load cell is proposed in the form of a cantilever with an electrical resistance strain gauge on the top and one on the under side, the force being applied at the free end of the cantilever (as in *Figure 3.18*). Design, produce, test and calibrate such a load cell for a specified force range (say 0 to 30 N) and accuracy (say 5%).

7. PRESSURE MEASURING SYSTEMS

Pressure is force per unit area and thus has the basic unit of N/m^2. This unit is given a name—the *pascal* (Pa).

$1 Pa = 1 N/m^2$

At the surface of the earth there is a pressure due to the atmosphere of about $100 kN/m^2$, i.e. 100 kPa. This pressure is sometimes referred to as a *bar*.

$1 bar = 100 kPa$

The atmospheric pressure varies from place to place and also depends on the weather. Many industrial pressures are however not measured relative to zero pressure, i.e. the *absolute pressure*, but relative to the prevailing atmospheric pressure. Such relative pressures are known as *gauge pressures*.

Force measuring systems and pressure measuring systems are, in many instances, based on essentially the same principle since pressure is force per unit area. A common method of determining a force is to compare it with a known force, as in the simple analytical balance where the gravitational force acting on a known mass is compared with the gravitational force acting on the unknown mass. A common method of determining pressure is to compare the unknown pressure with a known pressure, the pressure generally being that produced by a column of liquid. Instruments using this principle are known as *manometers*.

A column of liquid of height h and cross-sectional area A has a volume of hA. If the density of the liquid is ρ then the mass of the liquid is $hA\rho$. The gravitational force acting on the mass is $hA\rho g$, where g has the value of the acceleration due to gravity. The pressure at the base of the

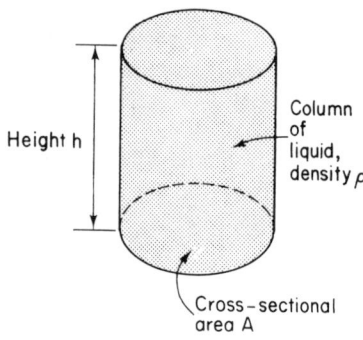

Height h

Column
of
liquid,
density ρ

Cross-sectional
area A

Figure 5.26

liquid column due to the liquid (*Figure 5.26*) is the gravitational force divided by the area of the column and thus is $hA\rho g/A = h\rho g$. The pressure is thus independent of the cross-sectional area of the liquid column.

The pressure acting on the top surface of the liquid column is the atmospheric pressure and thus the pressure at the base of the liquid column is equal to the atmospheric pressure plus $h\rho g$. The quantity $h\rho g$ represents the difference in pressure between the upper surface of the liquid and the base of the liquid column.

Figure 5.27a shows a *simple manometer*. The difference in the heights of the liquid columns in each arm of the U represents the difference in pressures at the two liquid surfaces, i.e. $h\rho g$ where h is the difference in height of the two liquid columns.

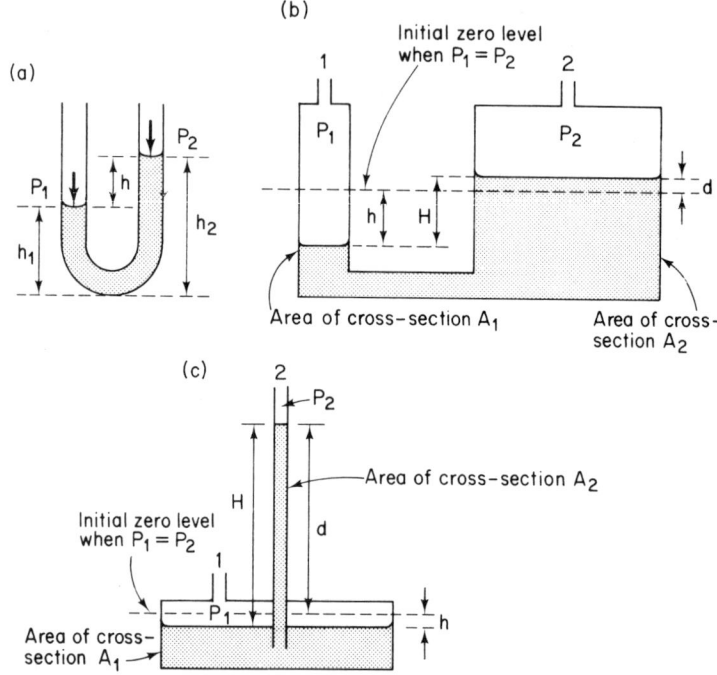

Figure 5.27 Manometers (a) A simple manometer (b) An industrial manometer (c) A cistern manometer

Pressure at base of column 1 $=\ h_1\rho g + P_1$

Pressure at base of column 2 $=\ h_2\rho g + P_2$

But the pressure at the base of column 1 must be the same as that at the base of column 2 as the two columns are connected together and any difference of pressure would result in movement of liquid from one liquid column to the other. Thus

$$h_1\rho g + P_1 = h_2\rho g + P_2$$

$$P_1 - P_2 = h_2\rho g - h_1\rho g = (h_2 - h_1)\rho g = h\rho g$$

Pressures measured by means of a manometer are sometimes expressed in terms of the difference in height of the liquid level, i.e. as for example 20 mm of mercury, rather than in units of pressure. As the density of mercury is about 13 600 kg/m³ and g about 9.81 m/s² the 20 mm of mercury, i.e. 0.020 m, exerts a pressure of $0.020 \times 13\,600 \times 9.81$ or about 2670 Pa.

The pressure differences that can be measured with a simple U-tube manometer are dictated by the differences in heights of liquid columns that can be accommodated in a reasonable size piece of equipment and the accuracy with which the difference in level can be read. With water as the liquid in the manometer a height difference of about 1 m is about the limit with the accuracy being of the order of a few millimetres, i.e. pressures up to about 10 000 Pa with an accuracy of about 20 Pa. With mercury as the liquid, pressure differences up to about 140 000 Pa can be measured with an accuracy of about 100 Pa. The pressures measurable with mercury are greater than those with water because of the greater density of mercury.

Figure 5.27b and *c* shows two industrial versions of the simple manometer. In both versions one limb of the U has a cross-sectional area considerably greater than the other limb. In the simple manometer both limbs of the U have the same size tubing. In *Figure 5.27c* the wider tube is wrapped round the narrower tube but it is still essentially a U-tube manometer. This type is known as a *cistern manometer*.

The pressure difference between the wide and narrow columns depends on the difference in height between the two columns and not the cross-sectional areas.

Pressure difference $= P_1 - P_2 = H\rho g$

But $H = h + d$, thus

$P_1 - P_2 = (h + d)\rho g$

Initially when P_1 was equal to P_2 the levels were the same in each limb, H was zero. When the two pressures became different liquid flowed from one limb to the other and so produced the difference in level H. The volume of liquid that left one limb must equal the volume of liquid that flowed into the other limb.

Volume of liquid leaving limb 1 $= A_1 h$

Volume of liquid entering limb 2 $= A_2 d$

Thus as these must be equal $A_1 h = A_2 d$

and hence $$h = \frac{A_2 d}{A_1}$$

Hence the pressure difference equation above can be written as

$$P_1 - P_2 = \left(\frac{A_2 d}{A_1} + d \right) \rho g$$

$$= \left(\frac{A_2}{A_1} + 1 \right) \rho d g$$

The movement of the level of the liquid in column 2 from its initial zero level, i.e. d, is thus proportional to the pressure difference and can be used as a measure of it. In many forms of industrial manometer this displacement d is determined by means of a float and lever system, as in *Figure 5.12*, though other forms of level measurement can be used. The level is measured rather than the difference in levels between two liquid

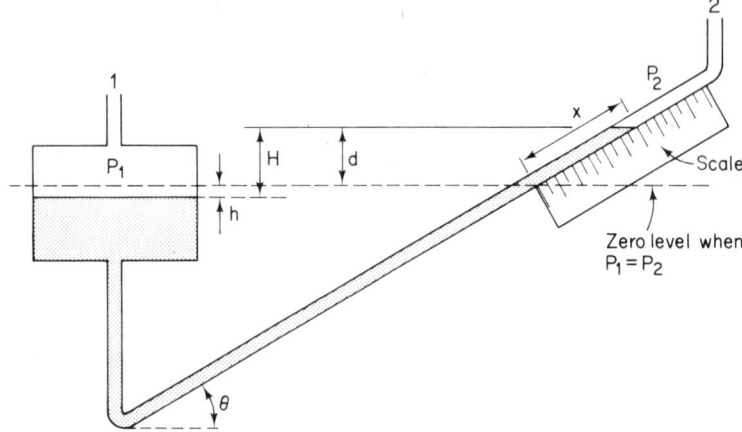

Figure 5.28 Inclined tube manometer

columns. This enables a greater accuracy to be achieved than with the simple manometer.

With the cistern manometer A_1 is so much greater than A_2 that the expression for the pressure difference can be approximated to

$$P_1 - P_2 = \rho dg$$

A special form of the U-tube manometer is the *inclined U-tube manometer*. (*Figure 5.28*). This has column 1 with a larger cross-sectional area than that of column 2 but additionally has column 2 inclined at some angle θ to the horizontal. Thus the vertical displacement d of the liquid level in column 2 is related to the movement x of the liquid along the inclined column by

$$\frac{d}{x} = \sin \theta$$

It is the displacement x that is measured rather than d. Thus the equation becomes

$$P_1 - P_2 = \left(\frac{A_2}{A_1} + 1 \right) \rho g x \sin \theta$$

and as A_2 is considerably smaller than A_1 the equation approximates to

$$P_1 - P_2 = \rho g x \sin \theta$$

If θ is 30° then as sin θ = 0.5 the equation becomes

$$P_1 - P_2 = 0.5 \, \rho g x$$

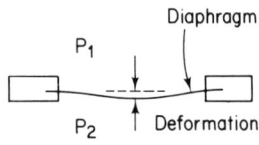

P_1-P_2 = pressure difference

Figure 5.29 Simple diaphragm as the basis of a pressure gauge

The displacement x is thus a direct measure of the pressure difference. Inclining the tube at 30° results in the displacement measured x being twice the displacement d that would have occurred with the tube vertical.

One way of measuring a force is to use a spring and relate the extension of the spring to the force causing the extension. The instrument utilising this principle is called the spring balance. A similar principle can be used as the basis of a pressure measuring system, i.e. an elastic deformation being used as a measure of the applied pressure.

The *diaphragm instrument* shown in *Figure 5.24* or that in *Figure 5.21* can be used for pressure measurement. A pressure difference between the two sides of a diaphragm (*Figure 5.29*) results in a deformation of the diaphragm. When the pressure difference changes the deformation of the diaphragm changes. The deformation can thus be used as a measure of the pressure difference. If the pressure on one side of the diaphragm is maintained constant then the deformation can be used as a measure of the pressure applied to the other side of the diaphragm. In *Figure 5.24* and in *Figure 5.21* different signal conditioners were used to magnify the diaphragm displacement and give a display.

The diaphragms shown in *Figures 5.29, 5.24* and *5.21* are all *plain diaphragms*, i.e. flat sheets of metal. The amount of movement of such a diaphragm is limited. Greater movement can be obtained with a *corrugated diaphragm (Figure 5.30a)*. Even greater movement can be obtained if two corrugated diaphragms are combined to give a *capsule*

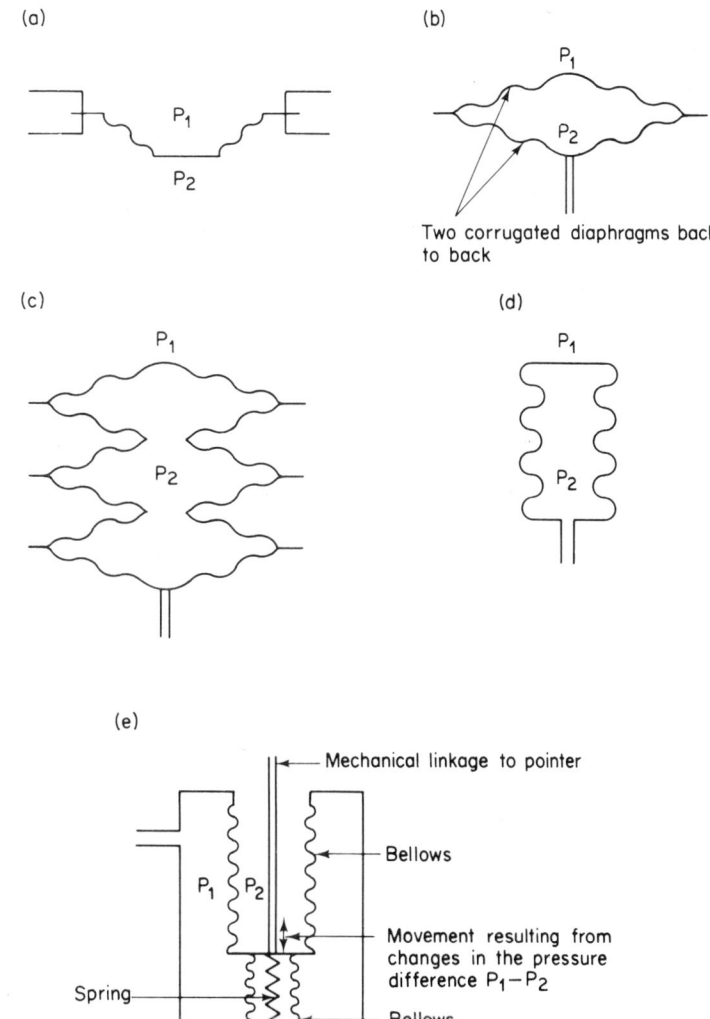

Figure 5.30 (a) A corrugated diaphragm (b) A capsule (c) A capsule stack (d) Bellows (e) A bellows gauge

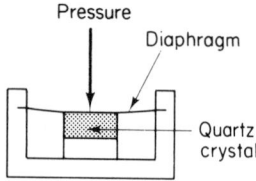

Figure 5.31 The basic principle of one form of a piezo-electric pressure gauge

(*Figure 5.30b*). The aneroid barometer shown in *Figure 3.6* uses such a capsule. A stack of capsules gives even greater sensitivity (*Figure 5.30c*). Instead of making up the capsule from a number of corrugated diaphragms a single piece of material can be used to give *bellows* (*Figure 5.30d*). *Figure 5.30e* shows the form of an industrial bellows gauge in which two sets of bellows are used and a spring. The use of the spring enables various sensitivities to be achieved from the same bellows design, changing the spring changes the sensitivity.

The *Bourdon tube gauge* (*Figure 1.1*) is a pressure gauge where an elastic deformation is produced as a result of pressure. The sensitivity of the Bourdon tube depends on the length of the tube, the thickness of the tube walls and the form of the cross-section of the tube.

Transient pressures can be measured with the diaphragm gauge employing strain gauges shown in *Figure 5.24*. The strain gauges respond quickly to any movement of the diaphragm. Metal diaphragms are capable of rapid response to pressure changes—manometers are not. Piezo-electric transducers are also used for transient pressure measurements; pressure applied to, say, a quartz crystal results in the production of a potential difference between opposite faces of the crystal. *Figure 5.31* shows the basic principle; pressure on the diaphragm results in the quartz crystal being compressed and a potential difference is produced.

For *calibration* of pressure measuring systems up to about 140 000 Pa a U-tube manometer can be used. The difference in height between the two columns of liquid needs to be measured accurately. *Figure 5.32a* shows a suitable form of manometer for calibration purposes. The height difference is always measured with reference to the same zero level, the mercury level being adjusted so that the mercury column in one limb is always at the same position. Coupled with a vernier scale

Table 5.8 Typical specification of a low pressure piezo-electric pressure transducer.

Pressure sensitivity > 400 pC/bar
Dynamic pressure range ± 5 bar
Static pressure without change in sensitivity ± 20 bar
Max. static overload pressure ± 100 bar
Natural frequency > 100 kHz
Frequency response (flat within ± 5%) 5 to 20 000 Hz
Temperature range, operational –54 to +700°C
Internal insulation resistance (at room temp.) >10^{10} ohm
Weight 270 g
Size (height × max. dia.) 46.4 × 60 mm
(For the significance of the natural frequency see Chapter 7)

Table 5.9 Typical example of a manufacturer's specification of a strain gauge pressure transducer.

Range:	20 to 700 bar, 0.7 to 10 bar.
Excitation:	10 V d.c. or a.c. r.m.s.
Full range output:	40 mV
Non-linearity and hysteresis:	± 0.5%.
Temperature range:	−54°C to +120°C when operating.
Thermal shift—zero:	0.030% full range output/°C
—sens:	0.030% full range output/°C
Weight:	142 g
Features:	Sealed construction. Fully immersible. Suitable for marine and general industrial applications.

---

this enables the accuracy of the simple manometer to be improved and serve as a standard. The other quantities that need to be known accurately are the density of the manometer fluid and the acceleration due to gravity at the position where the manometer is being used.

For calibration at higher pressures a *dead-weight tester* can be used. *Figure 5.32b* shows the basic form of the instrument. The pressure to be measured by the gauge under test is obtained by applying known masses to a piston of known area acting on the fluid in contact with the gauge. The masses are applied to the piston and then the plunger moved in until the fluid pressure in the chamber is just sufficient to lift the piston with its masses. When this happens the pressure in the chamber is the same as that exerted by the piston. This can be calculated as mg/A, where m is the mass acting on the piston of cross-sectional area A and g the local acceleration due to gravity. The mass is the mass of

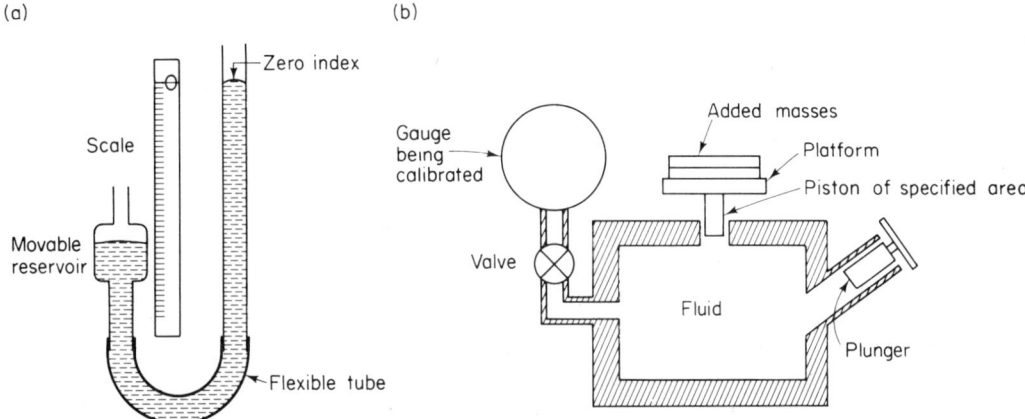

(a)   (b)

Figure 5.32 (a) Manometer (b) Dead-weight tester

both the masses on the piston table plus the mass of the piston and its table. Typical accuracies are of the order of 0.03% or less for a range of about 100 kPa to 30 000 kPa.

The advantages and disadvantages of the pressure measurement systems discussed in this section can be summarised as follows.

Simple manometer: With an appropriate liquid, can be used for pressure difference measurements of the order of 20 Pa to 140 kPa, generally not very accurate, cheap.

Cistern manometer: Same possible range as the simple manometer, can be made to give a display at a distance, not very accurate.

Inclined tube manometer: Same possible range as the simple manometer, more accurate than the simple manometer.

Diaphragm, capsule and bellows instruments: Pressure differences of the order of 1 kPa to 200 kPa can be measured depending on the form of the system, more rapid response to pressure changes than the manometers, more compact than a manometer.

Bourdon tube gauge: Pressure differences of the order of 10^3 Pa to 10^8 Pa can be measured depending on the form of the gauge, less sensitive than capsules or bellows, relatively cheap, robust.

Strain gauge transducer system: Pressure differences of the order of 10^3 Pa to 10^8 Pa can be measured depending on the form of the system, can be used for static or transient pressures.

Piezo-electric transducer systems: Can be used only for transient pressures, pressure changes of the order of 10^5 Pa to 10^8 Pa are commonly measured with such a system.

PROBLEMS

(7.1) A simple manometer with water as manometer fluid shows a difference in water levels in the two U limbs of 50 mm. What is the pressure difference in units of N/m^2? The density of water may be taken as 1000 kg/m^3.

(7.2) Explain the principle of operation of the cistern manometer.

(7.3) What advantages does the cistern manometer have over the simple U-tube manometer?

(7.4) A simple U-tube manometer is required for the measurement of a pressure difference of the order of 500 Pa. Specify a possible manometer size and liquid.

(7.5) Sketch the possible form of a direct reading cistern manometer, the display being shown by the movement of a pointer across a scale.

(7.6) *Table 5.9* gives the specification of a strain gauge pressure transducer. Explain the significance of each of the items in the specification.

(7.7) Under what circumstances would you advocate the use of a Bourdon tube gauge rather than any other pressure measuring system?

ASSIGNMENTS

(7.1) Use a dead-weight tester to calibrate a Bourdon tube gauge. Produce a calibration chart.

(7.2) Analyse the requirements for a pressure measuring system for some specific industrial situation, e.g. the air pressure line system, and propose a suitable system.

8. MEASUREMENT OF FLUID FLOW

When a fluid flows past a constriction in a pipe its velocity increases, reaching a maximum where the cross-section of the pipe is a minimum. The increase in velocity is at the expense of the pressure in the fluid and thus there is a pressure drop at the constriction. Beyond the constriction where the pipe returns to its original size the velocity drops almost back to its original value, the pressure rising almost back to its original value. The pressure difference between that in the fluid prior to the constriction and that at the constriction depends on the initial velocity of the fluid and both the cross-sectional area of the pipe prior to the constriction and that at the constriction. Thus a measure of the pressure difference can be used as a measure of the fluid velocity. This is the basic principle used in a number of forms of flow measuring system.

Figure 5.33 shows a number of forms of *constriction flowmeter*. *Figure 5.33a* shows the basis of the *venturi tube*. The pressure difference is measured between the tube prior to the constriction and in the tube at the constriction. In a simple form of the instrument the pressure difference may be measured by connecting a simple U-tube manometer between the two, one limb being connected to each point. *Figure*

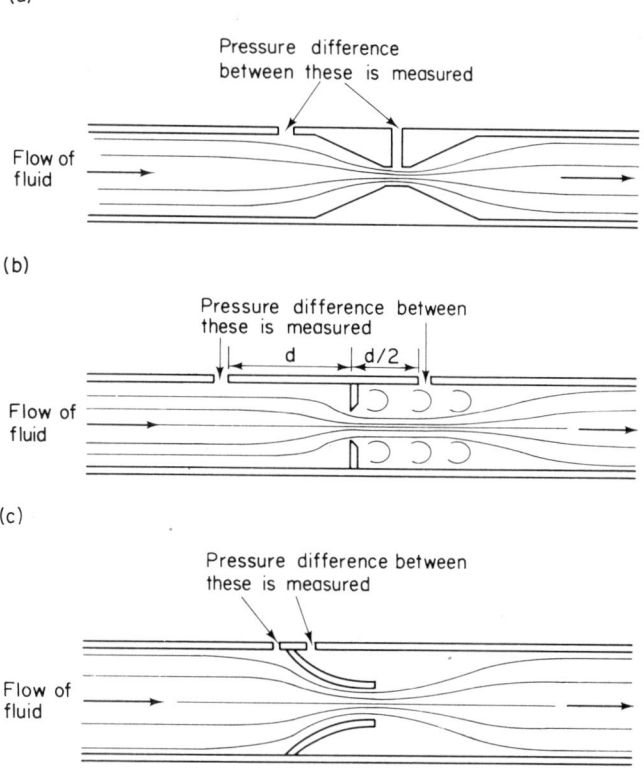

(a)

Pressure difference
between these is measured

Flow of
fluid

(b)

Pressure difference between
these is measured

d

d/2

Flow of
fluid

d

(c)

Pressure difference between
these is measured

Flow of
fluid

Figure 5.33 Constriction flowmeters (a) Venturi tube (b) Orifice plate (c) Nozzle

5.33b shows the *orifice plate flowmeter.* This is simply a disc with a hole through the middle. The result of placing this in the tube is to give a fluid flow pattern similar to that produced with the venturi tube and hence a pressure difference between the flow in the full width tube and that where fluid flow has become narrowest. The pressure difference is measured between a point a distance equal to the diameter of the tube upstream and a point a distance equal to half the diameter downstream. A *nozzle (Figure 5.33c)* can be used to give essentially the same results as with the venturi tube. Of the three forms of constriction flowmeters described above, the venturi tube offers the least resistance to the fluid flow and thus has least effect on the rate of flow of fluid through the pipe. The orifice plate offers the greatest resistance and hence its use results in the greatest energy losses from the fluid. The orifice plate is nowever the cheapest, the venturi tube being the most expensive.

The basic relationship between the pressure difference Δp and the rate of flow Q, in units of volume per second, is

$$Q = \frac{C_d a}{\sqrt{(1 - (a/A)^2)}} \sqrt{\frac{2\,\Delta p}{\rho}}$$

where a = cross-sectional area of pipe at the constriction,

(a)

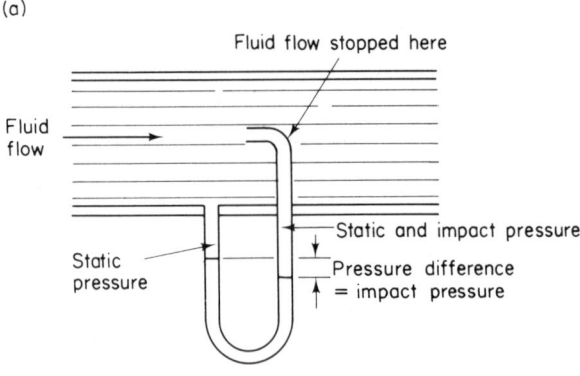

(b)

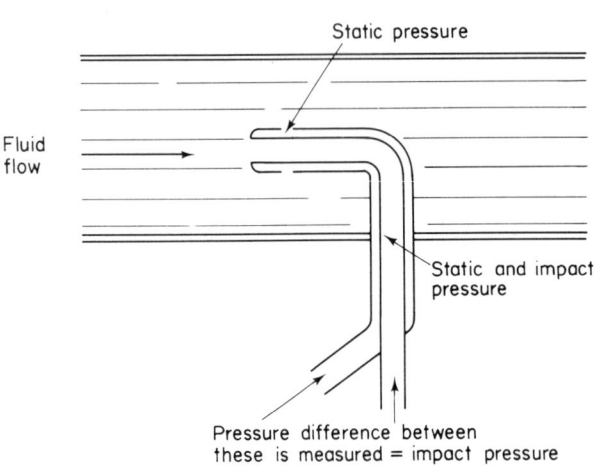

Figure 5.34 Pitot tube (a) Basic principle (b) Typical assembly

A = cross-section area of the pipe prior to the constriction,

ρ = density of the fluid,

C_d = a constant which depends on the form of the constriction and the fluid used.

For a venturi a typical value of C_d is about 0.97, for the orifice plate the value may be as low as 0.62. A point to notice is that for a particular flowmeter the rate of flow is proportional to the square root of the pressure difference.

There are a number of instruments which can be used to measure fluid velocity, the constriction flowmeters measuring the volume rate of flow. Probably the simplest of such instruments is the *Pitot tube* (*Figure 5.34*). A moving fluid has kinetic energy due to its moving with a velocity and a potential energy due to the fluid having a static pressure. If the fluid were stationary we would still be able to measure a pressure —the static pressure. In the Pitot tube the pressure of the fluid when 'stationary' is compared with the pressure due to both the kinetic plus static pressure, the pressure difference thus being that due to the movement of the fluid alone. The energy of the moving fluid is its potential energy plus its kinetic energy, i.e.

$$MgH + \tfrac{1}{2}MV^2$$

where M is the mass of an element of the fluid moving with a velocity V, g is the acceleration due to gravity and H the pressure head, i.e. height of a column of the liquid, due to the static pressure. This energy is all transformed into potential energy by the fluid suffering an impact with the Pitot tube and being brought to rest.

$$MgH + \tfrac{1}{2}MV^2 = Mgh$$

h is the pressure head resulting from the conversion of the total energy into potential energy, i.e. the movement of a liquid column in a manometer.

Rearranging the equation gives

$$V^2 = 2g(h - H)$$

The quantity $(h - H)$ is the difference in levels of the liquid in the manometer. This difference is thus proportional to the square of the fluid velocity.

Pitot tubes can be used for the measurement of fluid velocities as low as 1 m/s and as high as 60 m/s. Above this value corrections have to be made because of the compressibility of the fluid. Pitot tubes

(a)

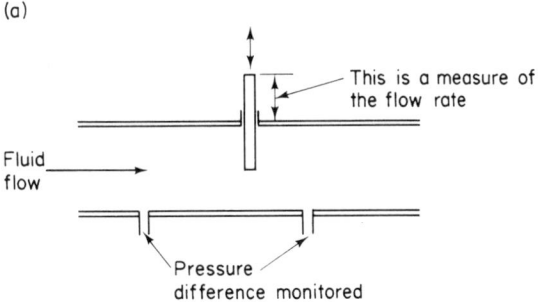

(b)

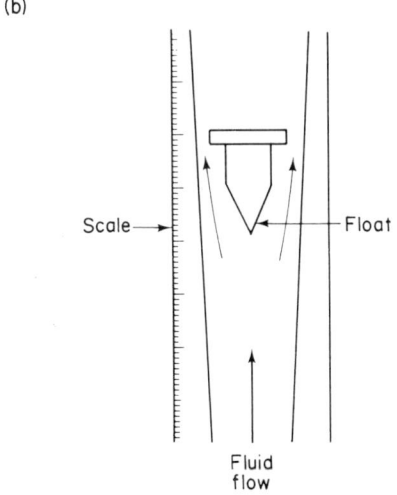

Figure 5.35 Variable area flowmeters (a) Variable area gate meter (b) Rotameter

coupled with Bourdon gauge tubes are used in aircraft for the measurement of air speed.

There are a number of forms of flowmeter called *variable area flowmeters*. These depend on the same principle as the constriction flowmeters but instead of measuring the pressure difference between the full width tube and the constriction the size of the constriction is varied to give a constant pressure difference. *Figure 5.35* shows two forms of such flowmeters. *Figure 5.35a* is essentially an orifice plate meter in which the size of the orifice is adjusted to give a constant pressure difference, the position of the gate which moves across the tube being a measure of the flow rate. The flowmeter is called a *variable area gate flowmeter*. Probably the most common variable area meter, is the *rotameter* (*Figure 5.35b*). This employs a float in a tapered vertical tube. The fluid in flowing past the float has to pass through a constriction, the gap between the float and the walls of the tube. This causes a pressure difference which results in the float being pushed up the vertical tube. The float will stop moving up the tube when the pressure difference is just sufficient to balance the weight of the float. As the tube is tapered the gap between the float and the tube walls increases as the float moves up the tube. The greater the flow rate the greater the pressure difference for a particular gap. Thus the float moves up the tube a height which depends on the rate of flow. A scale alongside the tube can thus be calibrated to read directly the flow rate appropriate for a particular height of float.

By a suitable choice of rotameter so fluid flow rates can be measured over quite a large range, from flow rates as low as about 30 ml/s to as high as 120 l/s.

1 litre (1) = 1 cubic decimetre (dm^3) = 10^{-3} m^3

The rotameter is a relatively cheap flowmeter capable of having a long life with little maintenance or recalibration being required.

Another group of flowmeters is known as *positive displacement meters* and are found typically in petrol pump meters, water meters and gas meters. All the forms depend upon some method of dividing up the flowing fluid into known volume packets and then counting the number of packets for the total volume flow or the number of packets per second for the flow rate. *Figure 5.36* shows one form of such a meter, the *rotating lobe meter*. For one complete rotation of the lobes a definite volume of fluid passes through the chamber. Each rotation can be used to operate a counter and thus the number indicated by the counter becomes a measure of the fluid volume that has flowed through the chamber. The meter is capable of high accuracy.

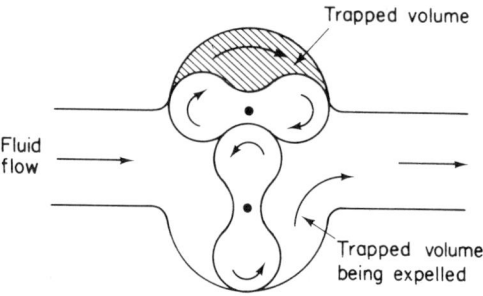

Figure 5.36   Rotating lobe meter

By supplying power to the lobes they can be rotated and so extract from a source a measured volume of fluid. The device then becomes a metering pump. This is the type of application found at the petrol pump when a measured volume of petrol is purchased.

Another form of flowmeter consists of a rotor or propeller that is supported centrally in the pipe along which the flow occurs and which rotates as a result of the fluid flow. Such a flowmeter is generally called a *turbine meter*. *Figure 5.37* shows one form of such a meter. The faster the fluid flows the faster the rotor rotates. The rate of revolution of the rotor can be determined by using an electromagnetic pick-up. This could be a small permanent magnet mounted at the tip of one the

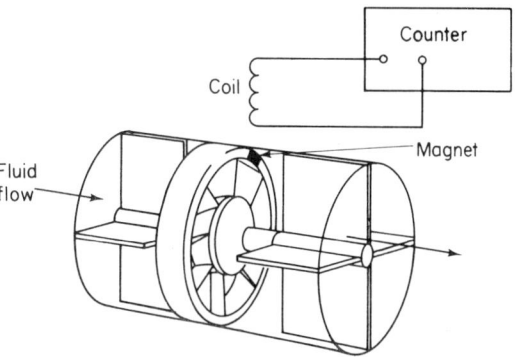

Figure 5.37 Turbine meter

rotor blades with a coil being placed just outside the tube. As the magnet moves past the coil it induces an e.m.f. in the coil. This voltage pulse can then be counted and so enables the number of revolutions of the rotor to be counted. The faster the fluid flow the greater the count per second. Some turbine meters are used to give only the total count and hence the total amount of fluid that has passed through a pipe.

There are many ways of *calibrating* flowmeters, the method often depending on the type of flowmeter concerned. One method is to calibrate the flowmeter against a more accurate flowmeter, the two

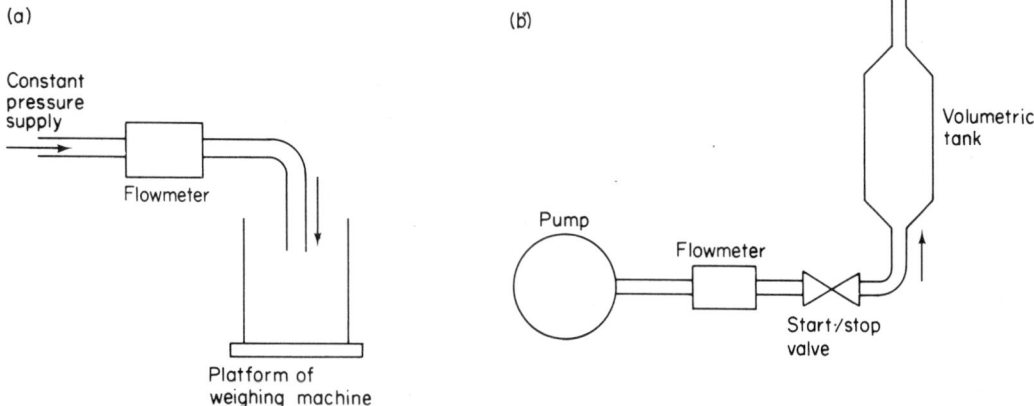

Figure 5.38 Calibration of flowmeters

being put in series with the same flow through each. Another method is to measure the volume of fluid passed through the flowmeter in a measured interval of time (*Figure 5.38a*). The flow rate needs to be maintained constant during the flow and thus a constant pressure supply is required. The volume is often deduced from a measurement of the mass with a balance. With the positive displacement meter where the total quantity passed through the meter is of concern the calibration is often achieved by pumping liquid through the meter and measuring the total volume passed in a measured time (*Figure 5.38b*).

The advantages and disadvantages of the different forms of fluid flow measurement systems discussed in this section can be summarised as follows.

Venturi tube: Can be used for both liquid and gas flow, measures volume rate of flow with reasonable accuracy, little pressure loss, long life without maintenance or recalibration.

Orifice plate flowmeter: Can be used for both liquid and gas flow, measures volume rate of flow with reasonable accuracy, a greater pressure loss than the venturi tube, long life without maintenance or recalibration, relatively cheap.

Nozzle flowmeter: Can be used for both liquid and gas flow, measures volume rate of flow with reasonable accuracy, less pressure loss than the orifice plate flowmeter but more than the venturi tube, long life without maintenance or recalibration, cheaper than the venturi but more expensive than the orifice plate flowmeter.

Pitot tube: Can be used for both liquid and gas flow, measures fluid velocity with reasonable accuracy, little pressure loss, cheap, not suitable for very small fluid velocities.

Variable area gate flowmeter: Can be used for both liquid and gas flow, measures volume rate of flow, can be made remote reading, produces a pressure loss similar to the orifice plate flowmeter.

Rotameter: Can be used for both liquid and gas flow, measures volume rate of flow, not highly accurate, produces a significant pressure loss, long life without maintenance and recalibration, cheap.

Positive displacement meter: Can be used for both liquid and gas flow, can be used to measure total quantity delivered or volume flow rate, accurate, produces a significant pressure loss, a reasonable life without maintenance or recalibration, not cheap.

Turbine meter: Can be used with liquids, can be used to measure total quantity delivered or volume flowrate, some pressure loss, not cheap.

PROBLEMS

(8.1) The orifice plate flowmeter is just another version of the venturi tube. Explain this statement.

(8.2) How do the results obtained from a pitot tube differ from those obtained with an orifice plate flowmeter?

(8.3) Explain the principle of the rotameter.

(8.4) Specify a system appropriate for the measurement of the total volume of liquid taken from a tank.

(8.5) A measuring system is required which can measure the rate of flow of a liquid along a pipe without there being any significant pressure loss. The rate of flow is of the order of 500 ml/s. Specify a possible system.

ASSIGNMENTS

(8.1) Calibrate a flowmeter.

(8.2) Petrol pumps deliver fuel through a hose into the tank of a car, the amount of fuel delivered being monitored. Analyse the situation and produce a specification for a system to measure and display the amount of fuel delivered. What accuracy is required? What rate will the system have to cope with? What is the best form of display?

9. TEMPERATURE MEASUREMENT SYSTEMS

Chapter 2 includes a consideration of the various types of transducer used for temperature measurement. These can be summarised as those depending on:

(1) A change in dimensions of material,
(2) A change in electrical resistance,
(3) Thermo-electricity,
(4) A change in the intensity and colour of the radiation emitted by a hot body,
(5) A change in state.

Probably the most common of the expansion type of thermometer is the *mercury-in-glass thermometer.* Other liquids can however be used. The range of a mercury-in-glass thermometer can be from about $-35°C$ to $+600°C$. Mercury boils under normal atmospheric pressure at $357°C$ and thus the higher range mercury-in-glass thermometers have nitrogen under pressure above the mercury surface. *Alcohol-in-glass thermometers* are used over the range $-80°C$ to $+70°C$, *pentane-in-glass thermometers* over the range $-200°C$ to $+30°C$. All such thermometers respond only slowly to temperature changes. They are not suitable for the measurement of the surface temperature of a solid. Thermometers classified as 'chemical' can be used as secondary standards in laboratories and may be supplied with a certificate giving the calibration. The less accurate thermometers are classified as 'industrial'.

Liquid-in-glass thermometers are not robust. A more robust thermometer employing the expansion of a liquid for the measurement of temperature is the *liquid-in-metal thermometer. Figure 5.39* shows one form of such a thermometer. The thermometer bulb is connected to a Bourdon tube and filled with the appropriate liquid. When the liquid expands the Bourdon tube straightens slightly and causes a pointer to

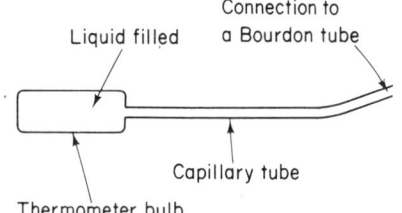

Figure 5.39 Liquid-in-metal thermometer

Table 5.10 Typical example of a catalogue entry for mercury-in-steel thermometers.

---

(1) Ranges covered lie between $-20°C$ and $530°C$, maximum error 1% of scale;
(2) linear scale;
(3) backlash avoided by direct drive to pointer;
(4) compensation for temperature effects in capillary tubing and case by use of microbore tubing and bimetallic links in the head;
(5) welded steel bulbs;
(6) capillary tubing of copper-protected drawn steel.

In addition, distance thermometers are not affected by distance of bulb vertically above or below dial and are effective up to distances of 40 m. Standard ranges available.

$-20°C$ to $30°C$; $-10°C$ to $40°C$; $-10°C$ to $120°C$; $-10°C$ to $170°C$; $-10°C$ to $420°C$; $0°C$ to $80°C$; $10°C$ to $60°C$; $10°C$ to $170°C$; $10°C$ to $300°C$; $20°C$ to $200°C$; $30°C$ to $120°C$; $30°C$ to $530°C$.

---

move across a scale. When mercury is used as the liquid the range is −39°C to +650°C, with alcohol the range is −46°C to +150°C. The accuracy of such instruments is of the order of ± 1% of the range used.

In the *constant volume gas thermometer* the change in pressure brought about by a change in temperature for a fixed volume of an inert gas is used to give a measure of the temperature. The thermometer can have much the same form as that in *Figure 5.39* but having a gas in place of the liquid.

The bulb of a gas filled thermometer is generally fairly large, about 50 to 100 cm³. The relationship between the pressure and the absolute temperature, for a constant mass at constant volume, is given by the ideal gas equation

$$\frac{P}{T} = \text{a constant}$$

where T is the absolute temperature, i.e. on the Kelvin scale.

The *vapour pressure thermometer* has essentially the same form as the liquid-in-metal thermometer (*Figure 5.39*). It operates on the

Table 5.11 Typical example of a catalogue entry for a vapour pressure thermometer.

---

(1)   Standard ranges between −15°C and 310°C, maximum error 1% of scale·
(2)   non-linear scales, intervals greater for higher temperatures;
(3)   steady movement of pointer ensured by a precision magnifying movement;
(4)   temperature shown is that of the bulb, the capillary and case being unaffected by changes;
(5)   bulbs generally of non-ferrous materials. Distance thermometers require special calibration where bulb is more than 2 m vertically above or below dial and are normally limited to a 65 m run.
Standard ranges available.
−15°C to 35°C; −15°C to 50°C; −15°C to 60°C; −15°C to 80°C; 20°C to 100°C; 20°C to 110°C; 20°C to 120°C; 25°C to 150°C; 25°C to 175°C; 80°C to 200°C; 80°C to 280°C; 80°C to 310°C.

---

principle that if the temperature increases the amount of a liquid that vaporises will increase. The bulb of the vapour pressure thermometer is thus only partially filled with liquid, the space above the liquid containing the vapour from the liquid. Typical liquids used are methyl chloride and ethyl alcohol. With methyl chloride the range of the thermometer is of the order of 0°C to 50°C, with ethyl alcohol the range is 90°C to 170°C. Typical accuracy is about ±1% of the scale range used.

One of the problems with the liquid-in-metal thermometer and the constant volume gas thermometer is that the temperature at the bulb and along the tube connecting the bulb to the gauge both have an effect on the temperature indicated by the gauge. Making the connecting tube a very fine bore is one way of reducing the error due to changes in temperature of the gas or liquid in the connecting tube. Another way is to use a second capillary tube alongside the main capillary tube and have it of the same length as the main capillary tube but terminating just before the bulb. It is connected to a second Bourdon tube. The display pointer is then driven by the difference in movement between the two Bourdon tubes, the difference being a measure of the temperature of just the bulb. *Figure 5.40* shows the type of arrangement.

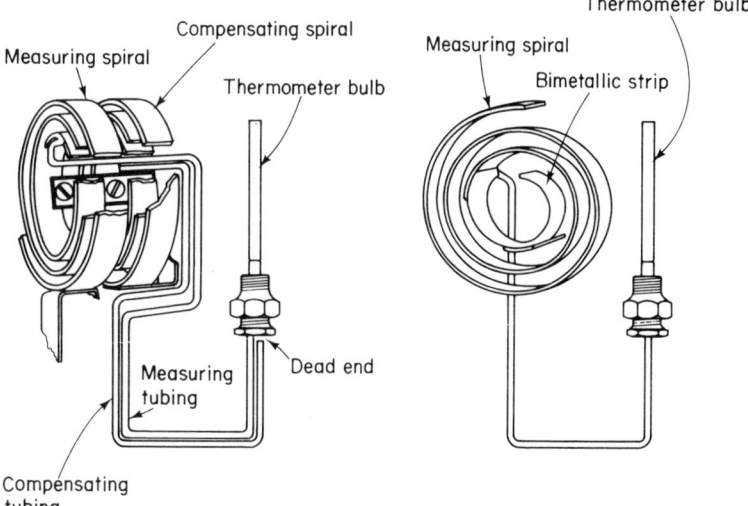

Figure 5.40
(a) Use of an extra capillary tube to compensate for temperature changes in the capillary tube connection to the bulb
(b) Use of a bimetallic strip to compensate for temperature changes of the Bourdon tube

Figure 2.3 shows the form of another instrument for temperature measurement based on expansion—the *bimetallic strip thermometer*. Such instruments are robust and cheap. They can be used within the range of about −30°C to 550°C with an accuracy of about ±1% of the scale range used.

A *resistance thermometer* generally depends on the measurement of the resistance of a metal coil of wire as an indicator of temperature. The measurement system has often a Wheatstone bridge as signal conditioner. *Figure 5.41* shows the basic form of the system. The coil of wire is connected by leads to one arm of the bridge. As the temperature

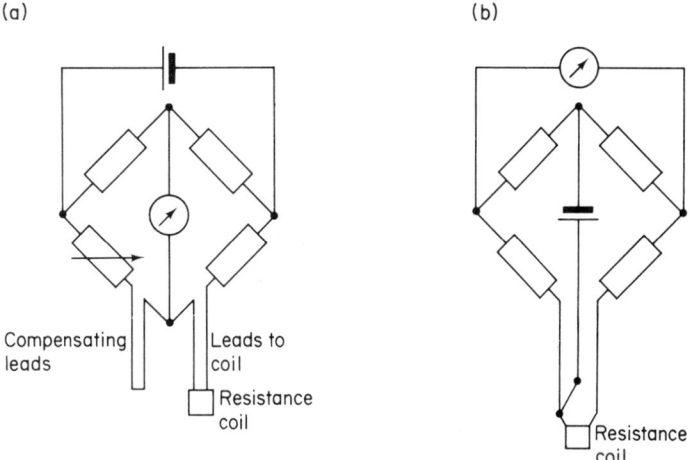

Figure 5.41 Reactance thermometer systems
(a) Two-lead resistance coil with compensating leads
(b) Another form of compensation using three leads to the resistance coil

required is only that at the coil and not that along the leads compensation has to be made for the effect of temperature changes on the resistance of the leads. This can be done by including a pair of compensating leads in another arm of the bridge. Any changes in resistance in the leads to the coil are cancelled by corresponding changes in resistance of the compensating leads. An alternative to this compensation is to use leads of very low resistance so that the effect of changes in lead resistance is considerably reduced. Electrical resistance thermometers are capable of high accuracy and stability and are thus often used as secondary standards. They can be used over the range $-260°C$ to $1000°C$, different metals being used for different parts of this range. Industrial forms of the thermometer are generally used over the range $-200°C$ to $600°C$ with an accuracy of generally under 1% of the scale range.

Thermistors are semiconductor resistance elements with a much larger variation of resistance with temperature than given by a metal resistance coil (see *Figure 2.5*). Thermistors can be in the form of small beads, rods and discs. Because of their small size they can be used for measuring temperatures over quite a small area. Their small size also means a small thermal capacity and hence a rapid response to temperature changes. For accurate work thermistors are generally used with a Wheatstone bridge. The temperature range possible is about $-250°C$ to $650°C$. Some thermistors may however be used over a very limited range, possibly as small as $10°C$.

Thermocouples are described in Chapter 2 and the basic form illustrated in *Figure 2.2*. Because the temperature indicated by a thermocouple system is only that of the junction between two dissimilar metals they can be used for the measurement of temperatures over a very small area. Their small size also means a small thermal capacity and hence a rapid response to temperature changes. Thermocouples are generally used with potentiometer circuits as signal conditioners. Depending on the materials used for the thermocouple wires so the range covered can vary, within about $-200°C$ to $1700°C$.

Instruments used for the measurement of temperatures above about $700°C$ are called *pyrometers*. *Figure 2.8* showed one form of pyrometer, the *disappearing filament pyrometer*. This instrument depends on the variation in the intensity of the radiation emitted by a hot object, the hotter an object the brighter the light emitted by it. *Figure 5.42* shows another form of pyrometer, the *total radiation pyrometer*. The radiation emitted by the hot body is focused on to the hot junction of a thermocouple or a small resistance element. The signal from this junction or element is then a measure of the temperature of the hot object, the hotter the object the greater the amount of radiation emitted by it.

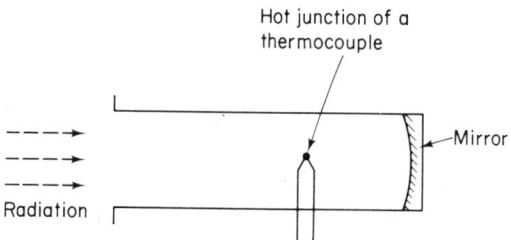

Figure 5.42 The basic form of a fixed focus total radiation pyrometer

The rate of radiation from a body at a temperature T, on the Kelvin scale, is given by

$$E = e\sigma T^4$$

where E is the total energy emitted per unit area per unit time, e is the emissivity, depending on the nature of the emitting surface and about 1 for a furnace and σ is a constant, called Stefan's constant and having a value of 5.7×10^{-8} W K^4 m^{-2}. Because of this fourth power relationship the characteristic of a total radiation pyrometer is non-linear. Total radiation pyrometers are generally used within the range 700°C to about 2000°C.

When a change of state occurs for solids there can be a change of shape or size or even a change in colour. These changes can be used as a measure of the temperature at which the change is occurring. *Pyrometric cones* are made of materials which at certain temperatures change shape, the temperature range available being between 600°C and 2000°C.

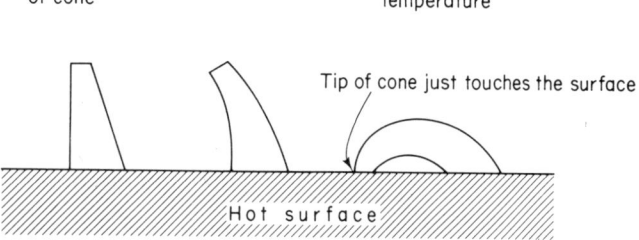

Figure 5.43 Pyrometric cones

Table 5.12 Typical example of range of paints (Thermocolor pigments).

Thermocolor No.	Original colour	Changed colour	Transition temperature in °C
		With one colour change	
1	Pink	Blue	40
2	Light green	Blue	60
2a	Pink	Blue	80
2b	Pink	Lilac	95
3	Yellow	Violet	110
4	Purple	Blue	140
4a	Blue green	Black	165
5	White	Brown	175
6	Green	Brown	220
7	Yellow	Red brown	290
8	White	Brown	340
9	Green	White	440
10	Red	Light grey	520
11	Red	Yellow	560
12	Yellow	Light green	640
13	Yellow	Olive green	715
13a	Beige	Brown	805
14	Grey	Dark brown	900
15	Green	Brown	1000
16	Light blue	Blue black	1100
17	Grey	Black brown	1200
17b	Light grey	Brown	1260
18	Grey yellow	Black	1350

Figure 5.43 shows the change in shape that occurs, the specified temperature being the one at which the cone tip has just bent over sufficiently to touch the hot surface. The accuracy is about ± 10°C.

Paints and crayons can be used to mark a hot surface, the colour of the paint or crayon mark changing colour at the specified temperature. Temperatures can be measured by this method within the range 40°C to about 1400°C and to an accuracy of about ± 5°C.

Temperature measuring systems can be *calibrated* against some standard temperature measuring system. The platinum resistance thermometer, calibrated by some authority, can be used as a secondary standard against which other temperature measuring systems can be calibrated. The primary standards used are however a number of fixed temperatures, the temperature scale given by these fixed points being known as the *International Practical Temperature Scale*. When a pure substance changes state, e.g. changes from solid to liquid, there is no change in temperature during the change. The graph of temperature against time during such a change is of the form shown in *Figure 5.44*. The temperature of the change of state is defined by the 'flat' portion of the graph. By specifying the temperature for such transitions the temperature scale is specified. The primary points of the scale are as follows. The term triple point refers to the situation where the solid, liquid and vapour phases of the substance are all in equilibrium.

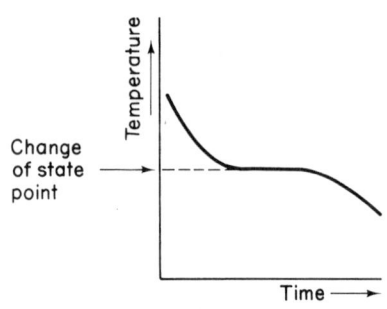

Figure 5.44 Change of state

Kelvin scale in K	Celsius scale in °C	
13.81	−259.34	Triple point of hydrogen
17.042	−256.108	Boiling point of hydrogen at a pressure of 33 330.6 Pa.
20.28	−252.87	Boiling point of hydrogen at standard atmospheric pressure.
27.102	−246.048	Boiling point of neon at standard atmospheric pressure.
54.361	−218.789	Triple point of oxygen.
90.188	−182.962	Boiling point of oxygen at standard atmospheric pressure.
273.16	0.01	Triple point of water.
373.15	100	Boiling point of water at standard atmospheric pressure.
692.73	419.58	Freezing point of zinc.
1235.08	961.93	Freezing point of silver.
1337.58	1064.43	Freezing point of gold.

As well as these primary points particular temperature measuring systems are specified for the intervals between the fixed points. Some additional points, secondary points, are also defined.

The advantages and disadvantages of the different forms of temperature measurement systems discussed in this section can be summarised as follows.

Liquid-in-glass thermometer: Direct reading, fragile, depending on liquid the range covered is −200°C to 600°C, capable of reasonable accuracy under standardised conditions, relatively cheap.

Liquid-in-metal thermometer: Robust, gives a display at a distance, depending on liquid the range covered is about −50°C to 650°C, no power supply required, direct reading, accuracy about ±1%.

Constant volume gas thermometer (industrial form): Robust, gives a display at a distance, range about −100°C to 450°C, no power supply required, direct reading, accuracy about ±1%.

Vapour pressure thermometer: Robust, gives a display at a distance, depending on the liquid used the range covered is about -50°C to 300°C, no power supply required, direct reading, accuracy about $\pm 1\%$.

Bimetallic strip thermometer: Robust, direct reading, general range about -30°C to 600°C, no power supply required, accuracy about $\pm 1\%$, relatively cheap.

Metal resistance thermometer: Generally requires a bridge to be balanced, requires a power supply, can have a reasonably distant display, general range -260°C to 1000°C, capable of high accuracy, can be used as a secondary standard.

Thermistors: Generally requires a bridge to be balanced, requires a power supply, can have a reasonably distant display, general range -250°C to 650°C, measures temperature over a small area, can respond quickly, tendency for calibration to change with time.

Thermocouples: Generally requires a potentiometer to be balanced though can be made direct reading, can have a reasonably distant display, general range -200°C to 1700°C though particular range depends on the wires used, measures temperature over a very small area, can respond quickly.

Disappearing filament pyrometer: Robust, no direct contact with the hot object, requires a power supply, general range 600°C to 1800°C though higher temperatures are possible, accuracy about $\pm 10^\circ$C.

Total radiation pyrometer: Robust, no direct contact with the hot object, general range about 700°C to 2000°C, accuracy about $\pm 10^\circ$C.

Pyrometric cones: Cheap, can be used over the range 600°C to 2000°C, accuracy about $\pm 10^\circ$C.

Paints and crayons: Cheap, can be used over the range 40°C to 1400°C, accuracy about $\pm 10^\circ$C.

PROBLEMS

(9.1) Compare the specifications for the mercury-in-steel and the vapour pressure thermometers given in *Tables 5.10* and *5.11*. What are the main differences between the two thermometers?

(9.2) Explain the compensation entry in the specification for the mercury-in-steel thermometer, *Table 5.10*.

(9.3) Explain the three lead form of compensation for a resistance thermometer.

(9.4) Explain how pyrometric cones are used for the measurement of temperature.

(9.5) *Table 5.12* lists the range of Thermocolor paints available when each paint is used to indicate just one temperature. Explain how the paints are used and specify the paints that would be needed for temperature measurements in the range 200°C to 350°C. What temperatures in this range can be measured?

(9.6) Which form of measuring system would be appropriate for the measurement of the surface temperature of a sheet of metal, the temperature being of the order of 200°C?

ASSIGNMENTS

(9.1) Assemble a thermocouple from two appropriate metal wires and use it with a sensitive galvanometer as indicator. Calibrate the system.

(9.2) Analyse the measuring system requirements of some industrial process which requires temperature measurement, e.g. the temperature of a furnace, and write a specification for the system.

6 Control systems

Aims: At the end of the chapter you should be able to:
Explain how feedback can be used to give automatic control systems.
Explain the basic terms associated with control systems.
Analyse and propose simple control systems.

The level of the chapter is one of introduction or appreciation rather than a detailed analytical treatment of control.

A FEEDBACK SYSTEM

Consider the problem of a person trying to walk along a straight line. If the person has his eyes closed and no-one conveys any information about his progress to him then though he may start off walking on the line it is extremely unlikely that he will continue walking along the line. If however he is given instructions, such as 'more to the left' or 'more to the right', then though he may zig-zag a bit about the line his progress will essentially be along the line, even though his eyes are closed. If he walks with his eyes open then he can see if he starts to move off the line and correct his walk so that he remains moving along the line. Whether the information about his position relative to the line is communicated to him by words or by sight there is what is called a *feedback* signal.

Figure 6.1 illustrates the above systems. In *Figure 6.1a* the person receives no information about his position relative to the line. The situation is said to be one of *open-loop control*. This is a process of trying to achieve a goal, in this case walk along a straight line, without any corrections being made en route. *Figure 6.1b* shows the situation where the person receives spoken information about his position relative to the line. This situation is said to be one of *closed-loop control*. This is the process of trying to achieve a goal with corrections being made en route as a result of information received about how the event is proceeding. *Figure 6.1c* shows the situation where the person is trying to walk along the line with his eyes open. This is also a closed-loop control situation. In closed-loop control there is a feedback of information so that the course of events can be modified.

The loop referred to above is the path of information and the action involved. In *Figure 6.1a* there is no closed loop, no return of information to the initial action. In *Figure 6.1b*, *6.1c* the loop is closed, information is fed back to modify the initial action. The modified action then results in further information being fed back which results in further modification to the action, which results in yet further information being fed back, and so on. The information-action loop is closed. In an open-loop control system the input to the system is not modified as a result of any information resulting from the action.

An electric fire is an open-loop control system (*Fgure 6.2*). The input signal is the decision to switch on the fire; this action results in an electric power input to the electric fire which then has as its output a temperature change in the room. The temperature change is not affected

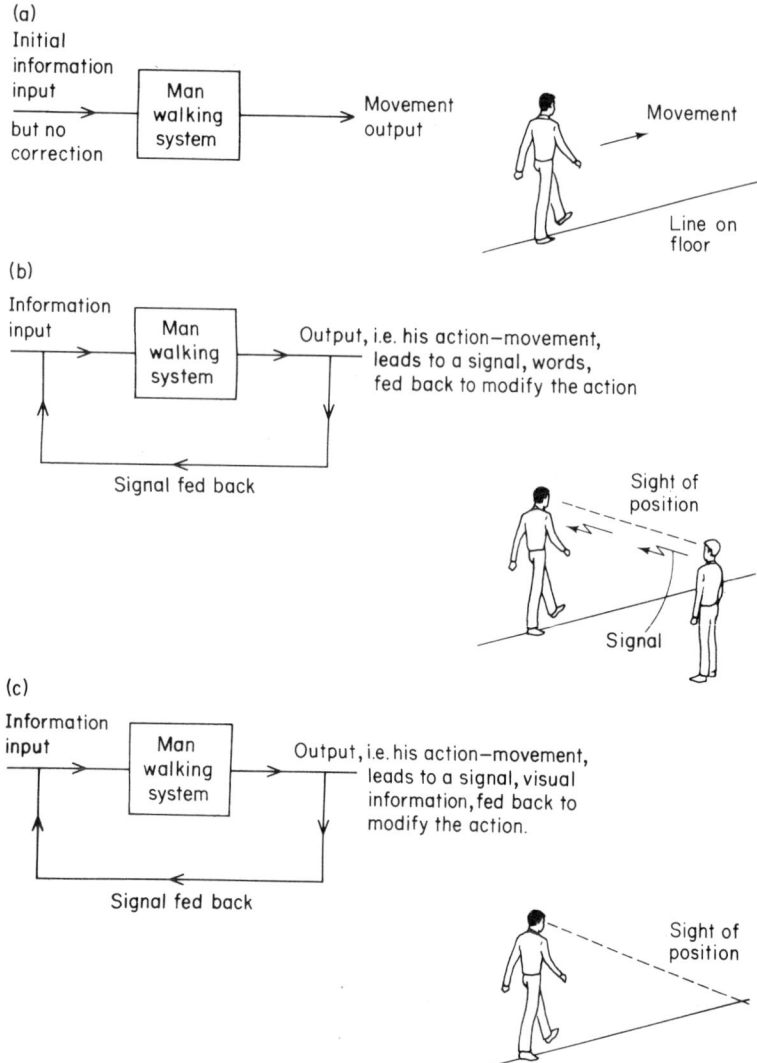

Figure 6.1 (a) Man walking with eyes closed and no correction input from any source. (b) Man walking with eyes closed and receiving correction information. (c) Man walking with eyes open.

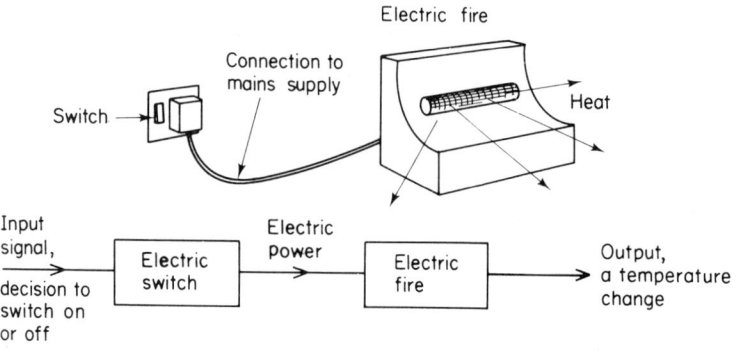

Figure 6.2   An open-loop control system, an electric fire

by any factor other than that imposed by the limitations on the input—
the amount of power that can be supplied to the fire. The result is a
controlled output because the power input is reasonably constant. If,
however, the room temperature changes because the sun shines through
the window then the output from the fire will not be modified, there is
no feedback to modify the input.

An open-loop control system can in general be represented by *Figure
6.3*. There is an input, the *demand signal*, to the control element. This
controls the input to the plant.

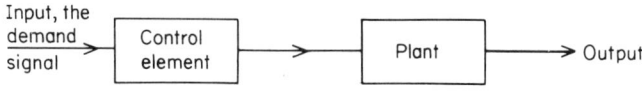

Figure 6.3 An open-loop system in general

(a)

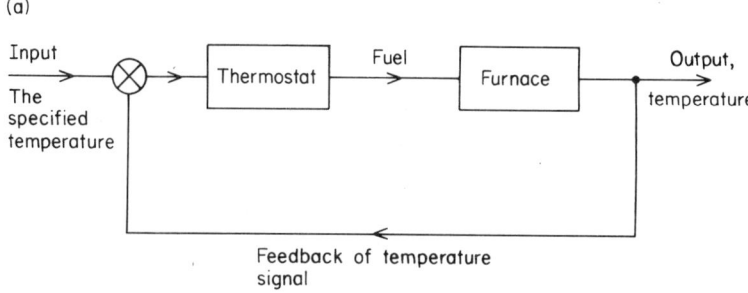

(b)

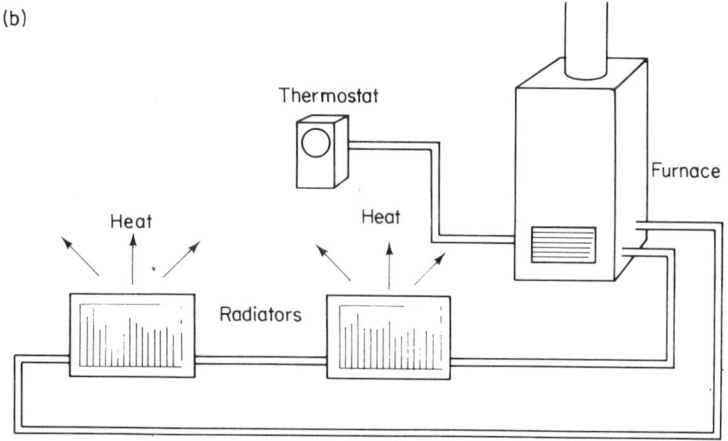

Figure 6.4 A closed-loop control system, central heating

A central heating system is generally a closed-loop control system
(*Figure 6.4*). The room thermostat is set to the required temperature
and this controls the fuel input to the furnace. The furnace output is a
temperature change for the room which is then fed back to the thermostat and so controls the input to the furnace. When the room temperature drops the thermostat switches on the fuel flow to the furnace. This
results in a heat flow into the room and a rise in temperature. When the
temperature rises to the value set for the thermostat it switches off

the fuel flow. The thermostat is an *on-off control*. The feedback is correcting automatically for changes in the system. If the sun shines into the room then the thermostat takes this into account and modifies the input to the furnace to compensate for it. The system is an *automatic control system*.

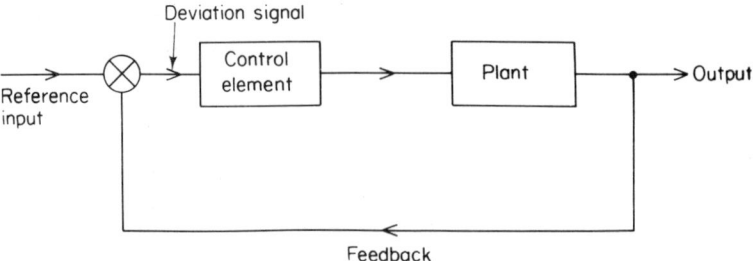

Figure 6.5 A closed-loop control system in general

A closed-loop control system can in general be represented by *Figure 6.5*. There is a *reference input* to the control element. This controls the input to the plant. There is feedback from the output back to the input to the control element. In the case of the thermostat-controlled furnace the feedback subtracts from the reference input, the specified temperature, and the *deviation signal* then is the input to the control element. The deviation signal might in some instances be referred to as the *error signal*, the difference between the feedback signal and the required

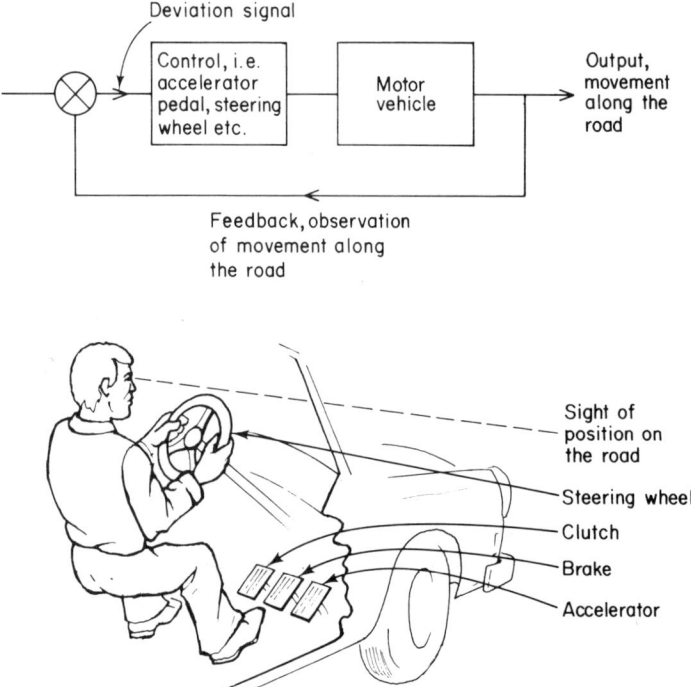

Figure 6.6 A motor vehicle control system

signal, in this case the required temperature. Where the feedback signal subtracts from the reference signal the feedback is said to be *negative feedback*. If the feedback signal adds to the reference signal the feedback is said to be *positive feedback*.

In the case of the feedback system described in *Figure 6.1b* the feedback to the man walking along the line is given by some other person watching and talking to him. Thus if the man strays away from the line the instruction would be words which direct him back towards the line, i.e. a signal to reduce the error, if the feedback is negative feedback. If however the instructions are words that direct him further away from the line, i.e. making the error larger, then the feedback is positive feedback.

The thermostat-controlled furnace, *Figure 6.4*, is a control system where the control is only intermittently exercised. The thermostat switches either on or off. The man walking along the line with some person issuing instructions can be a *continuous control* system if the man is continually correcting his path, as a result of instructions continually given, to minimise the error.

A motorist has a control system (*Figure 6.6*) in which the control is exercised by the positions of the accelerator pedal, the steering wheel, the brake pedal and the clutch pedal. The feedback loop governing the motion of the car along a road is the driver's observation of the position of the car relative to the road. The feedback is negative feedback, at least under normal circumstances, in that the feedback reduces the error. The control is also, generally, continuous control with the driver continually correcting the path and speed of the car.

NEGATIVE FEEDBACK– INSTABILITY

Negative feedback occurs where the feedback signal subtracts from the reference signal. The car control system described in *Figure 6.6* is a negative control system, the thermostat-controlled furnace in *Figure 6.4* is another negative control system. In both cases the output results in a feedback signal which, when subtracted from the reference signal, results in a change in the input to the plant in order to maintain the output at the required value. But the control of the car may lead to the car not moving perfectly down the road in the prescribed manner— after all some cars do run off the road despite the driver exerting control. The room temperature may not be constant despite the control of the furnace. This type of problem with a control system often arises when there is some *delay* in the feedback signal and the control element changing the input to the plant.

Consider the car control situation. The driver looks through the windscreen and sees the position of the car on the road. Assume the car to be off the path required. The driver sees this error and then reacts, say, by turning the steering wheel. But there is a time needed for the driver to see the deviation from the path and react (see Chapter 5 section 1 for a discussion of reaction time). This time might be a few tenths of a second or longer depending on the condition of the driver! During this reaction time the car is continuing at speed in the direction which is producing a deviation from the required path. If the reaction time is too long he might go off the road before the path is controlled. However assuming he does not go off the road then the correction given by the feedback moves the car back towards the correct path. When the correct path is reached the feedback signal should stop the correction being applied. The driver, seeing that he is in the right position on the road reacts by adjusting the position of the steering wheel. But as this operation takes time he will have moved past the correct path before he reacts. The result is that the car overshoots the correct path position. The behaviour of the system might thus look like that shown in *Figure 6.7*, the position of the car oscillating about the correct path position but never achieving exactly the right path. This type of behaviour is called *hunting*. The system has *instability*. In the case of the car control system the above is obviously a considerable simplification of the actual control process.

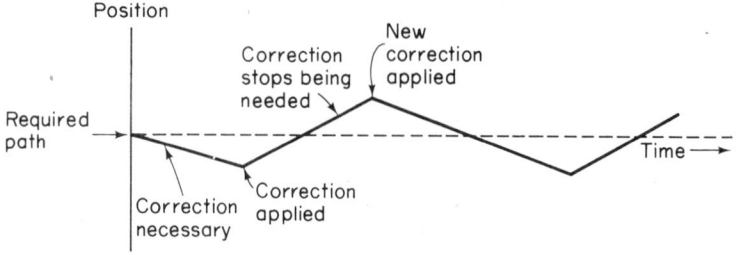

Figure 6.7 Hunting occurring with the motor control system

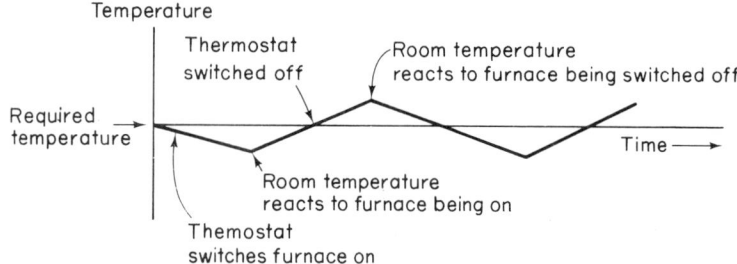

Figure 6.8 Hunting occurring with the thermostat-controlled furnace system

Figure 6.8 shows a similar situation with the thermostatically controlled furnace. The temperature oscillates about the required temperature. The delay in this case will probably be due to the time taken for the temperature of the water in the radiators to change.

PROCESS CONTROL SYSTEMS

A *process control system* is a system designed with the purpose of holding an output steady at the reference input value. *Figure 6.9* shows a simple system for the control of the level of a liquid in a tank. The required input value is a constant level. The operator observes the liquid level through a gauge glass and opens or closes a valve depending on the difference he sees between the observed level and the required level, some mark on the gauge glass. The system thus has a detecting, i.e. transducer, element together with some measuring element followed by a comparison element and finally a correcting or regulator unit operating on the plant. The detecting and measuring element is the gauge glass, the comparison element the difference in level between the observed level and the marked level on the glass as seen by the operator and the correcting unit is the valve operated by that operator. The plant operated on is the container of water.

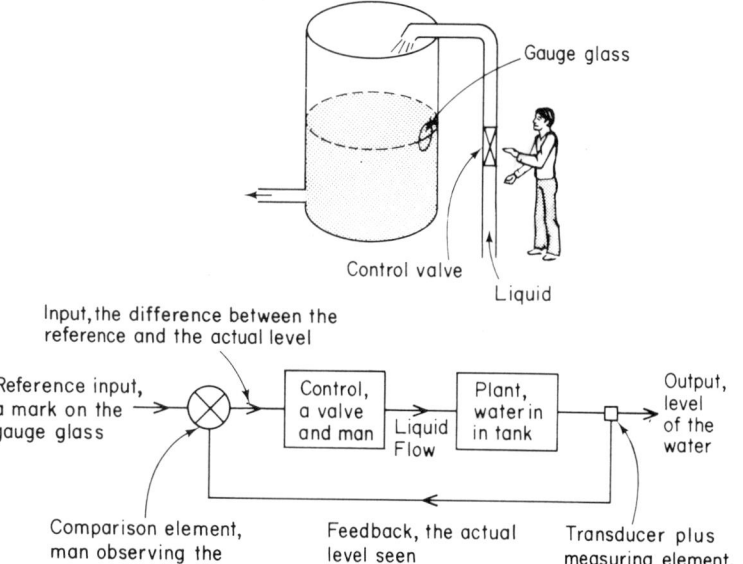

Figure 6.9 The control of water level in a tank

A process control system can thus be summarised as follows:

(1) A *transducer* and *measuring element*. These detect the changes in the condition being controlled and provide a signal which can be fed back to provide negative feedback.

(2) A *comparison element*. This compares the feedback signal with the reference input and provides a signal for the next stage according to the difference between the fed back signal and the reference.

(3) A *correcting unit*. This produces the change in the controlled condition, having its input from the comparison element.

(4) The *plant*. This is the system which produces the output being controlled. The correcting unit acts on this system so as to attempt to provide a constant output.

Figure 6.10 shows an automatic form of liquid level control. You probably have such a level control system for the cold water tank in your home. The reference is the initial setting of the ball-lever arm arrangment so that it just cuts off the water supply at the required level. When the ball, a float, is at a lower position, because the water level is below that required, there is an input of water into the tank. This flow continues until the ball has moved up to a position where the lever cuts off the water supply.

The thermostat-controlled furnace, *Figure 6.4*, is designed to give as the output a constant room temperature, the temperature being that specified by the reference input. Thus the thermostat might be set at 20°C and the control system expected to maintain the room temperature at that value. *Figure 6.11* shows the detail of a simple form of *thermostat*. When the temperature changes so the position of the bimetallic strip changes, opening or closing the contacts. Thus when the temperature drops below the reference value then the strip bends so as to close the contacts and switch on the furnace. If the temperature rises above the required value then the strip straightens so as to open the contacts and switch off the furnace.

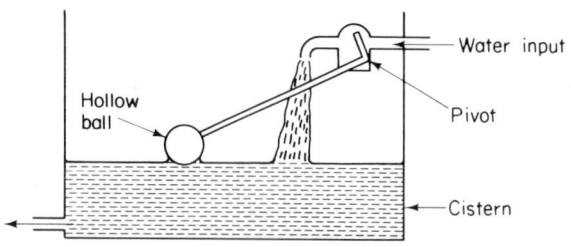

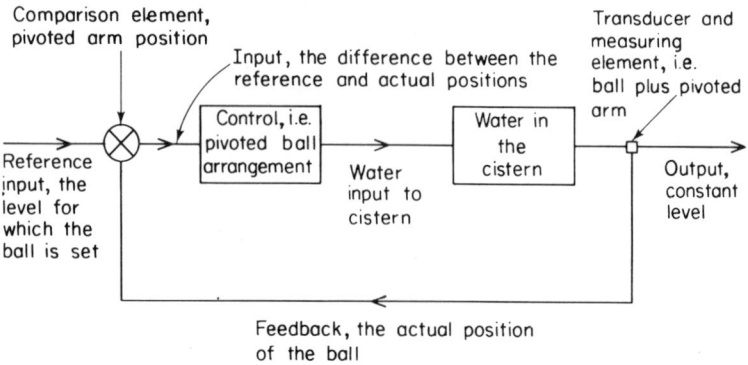

Figure 6.10 The control of water level in a cistern

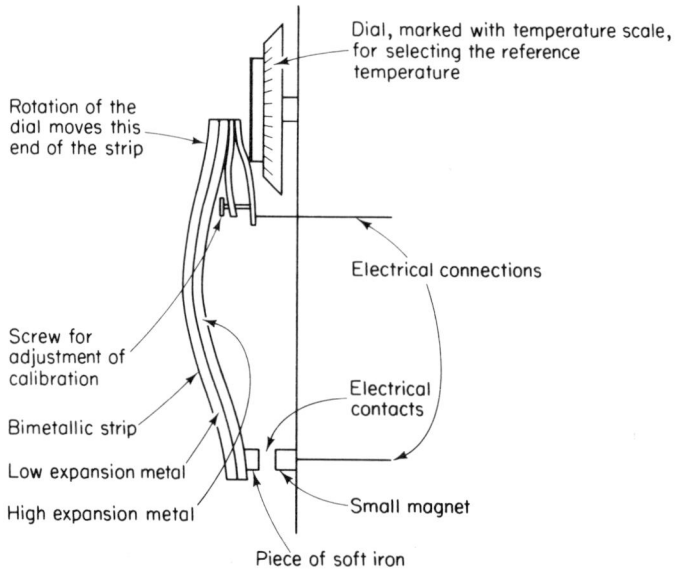

Figure 6.11 A thermostat

The thermostat controlled furnace is a process control system with the condition under control being the room temperature. The transducer and measuring element is the bimetallic strip. The comparison element is the electrical contact arrangement by which a signal, an electric current, is produced when the difference between the set position of the contacts and the actual positions are such as to close the contacts. The thermostat is an on-off control and does not provide a signal which is proportional to the difference between the required and observed temperatures but only when the observed temperature is below the required temperature. The signal from the thermostat is fed to some correcting unit which controls the input of fuel to the furnace.

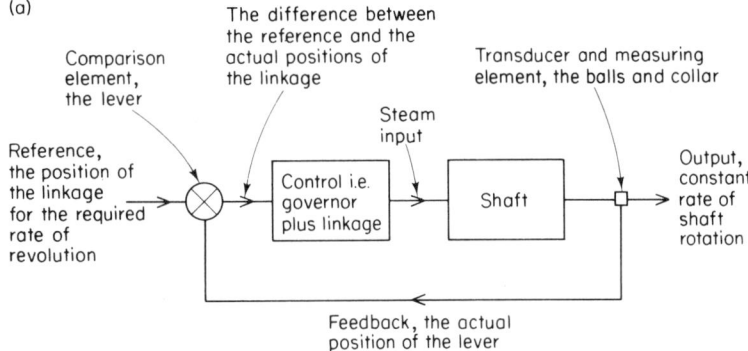

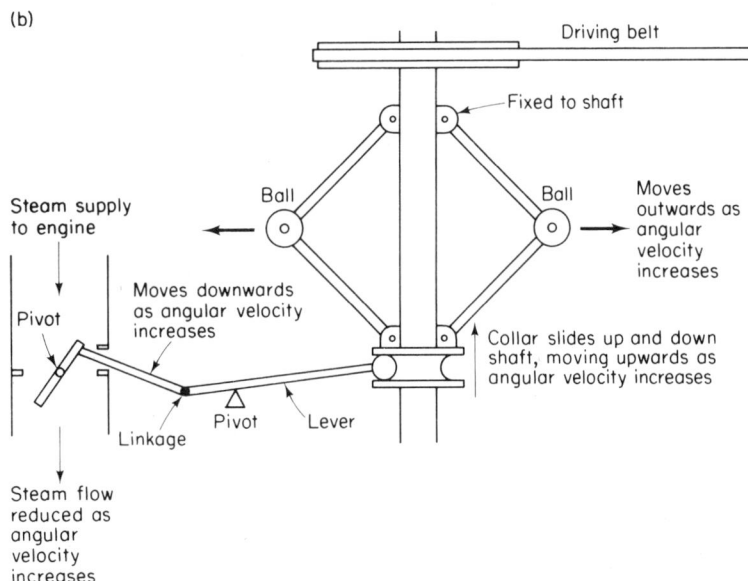

Figure 6.12 The Watt governer for controlling the rate of revolution of a shaft

Figure 6.12 shows a process control system used for the control of the rate of revolution of a shaft. The control element is the *Watt governer* described in Chapter 5, section 2 (*Figure 5.3*). When the rate of revolution of the shaft increases the balls move outwards which results in the collar being pulled up the shaft, causing movement of the linkage which has as its result the movement of the flap in the steam pipe in such a direction as to reduce the flow of steam. The reduction in the flow of steam to the engine results in less power being produced and so a reduction in the rate of revolution of the shaft. The shaft rate of revolution is thus controlled to some value pre-specified by the position of the linkage controlling the flap in the steam pipe.

The transducer and measuring element in the Watt governer is the balls and collar arrangement which results in a movement of a lever. The lever is the indicator of the movement. It is also the comparison element in that its movement at the linkage end results in an input to the correcting unit, the pivoted flap. The reference input for the system can be changed by raising or lowering the position of the pivot for the lever.

CONTROL VALVES

There are many process control systems where the input to the correcting unit is a pressure signal. *Figure 6.13* shows the first part of a *control valve*, the entire valve being the correcting unit. This part of the valve is known as the *actuator* and converts the pressure signal into a force and so movement of a shaft. The basis of the actuator is a flexible diaphragm,

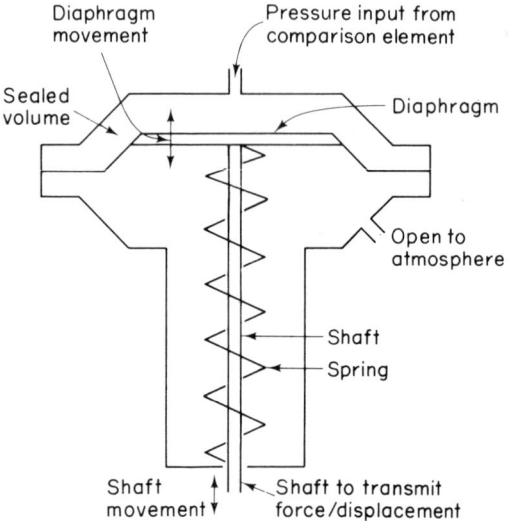

Figure 6.13 An actuator

one side of which is at the pressure determined by the comparison element and the other side being at atmospheric pressure. The diaphragm will move when there is a pressure difference btween the two sides, the diaphragm moving against a spring. When the diaphragm moves so does the shaft which is connected to the diaphragm. The movement of the shaft can then be used to open or close a valve.

If $p_1 - p_2$ is the pressure difference across the diaphragm and A is the area of the diaphragm then as pressure is force per unit area the force F acting on the diaphragm, and hence the spring is

$$F = (p_1 - p_2)A$$

The larger the area of the diaphragm the greater the force for a given pressure difference. The actuators are commonly operated by pressures in the region 20 to 100 kPa. The force when applied to the spring, the lower end of which is fixed, will cause the spring to become compressed. If the amount of compression x is proportional to the force, then

$$F = kx$$

where k is the force constant for the spring. Thus x is related to the pressure difference by

$$x = \frac{A}{k}(p_1 - p_2)$$

and so the amount by which the spring moves is directly proportional to the pressure difference and hence the signal from the comparison element. In practice this proportionality is achieved within an accuracy of a few per cent.

Figure 6.14 shows one possible form of a control valve, the actuator being the first part of the valve. The pressure change transmitted from the comparison element causes the shaft to move and so partially or

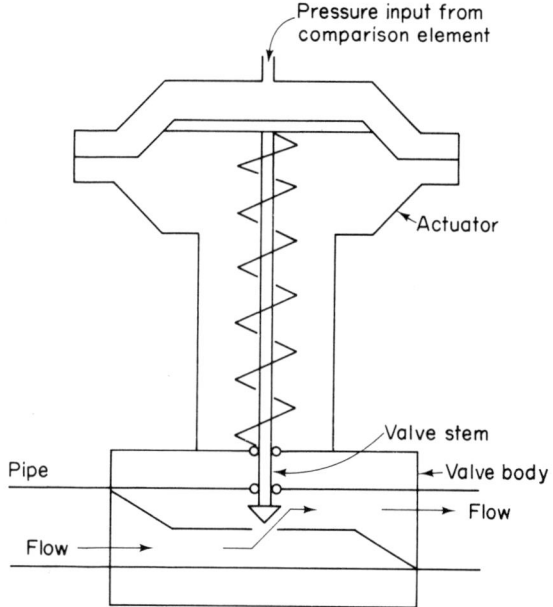

Figure 6.14 A control valve

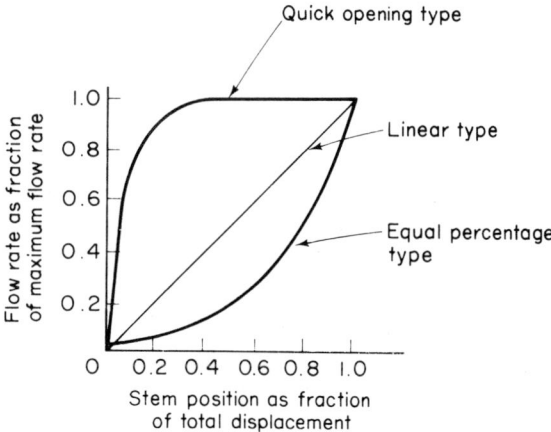

Figure 6.15 The three types of control valve characteristics

completely block the flow in the pipe. Thus the fluid flow along the pipe is controlled.

The performance of a valve will depend on the form and type of plug at the end of the shaft and how this reduces the orifice area through which the fluid flows. *Figure 6.15* shows the ways in which the shaft movement affects the flow for the three main types of control valves. The *quick opening type* gives a large change in flow rate for a small initial change in the position of the valve stem, i.e. the actuator shaft. This type of valve is often used to given an on-off control. The *linear type* gives a flow rate directly proportional to the amount by which the valve stem has been displaced. Thus the flow rate varies from zero to the maximum flow rate in accord with the position of the valve stem as it moves from its initial position to its maximum position. To have a flow rate of half the maximum flow rate the valve stem must be at its half way position. Thus

$$\frac{\text{Change in flow rate}}{\text{Maximum flow rate}} \propto \frac{\text{change in stem displacement}}{\text{maximum stem displacement}}$$

The third type of control valve is known as the *equal percentage type*. For each unit displacement of the valve stem the flow rate changes by the same percentage. Thus, for example, for the valve stem moving from its initial position by 5 mm the flow rate may change by 20% from 50 m/s to 40 m/s. When the valve stem moves by a further 5 mm the flow rate changes by a further 20%, from 40 m/s to 32 m/s. When the valve stem moves by a further 5 mm the flow rate changes by a further 20%, from 32 m/s to about 26 m/s.

$$\frac{\text{Change in flow rate}}{\text{Flow rate}} \propto \text{change in stem displacement}$$

This type of characteristic is an exponential.

Because of the widespread use of control valves which are actuated by a pressure signal from the comparison unit there is a need in many control systems to include a signal conditioner element between either the comparison unit and the control valve or between the transducer and the comparison unit. The signal conditioner has as its function the changing of a signal in one form to an equivalent signal in the form of a pressure change. *Figure 6.16* shows one form of signal conditioner, the

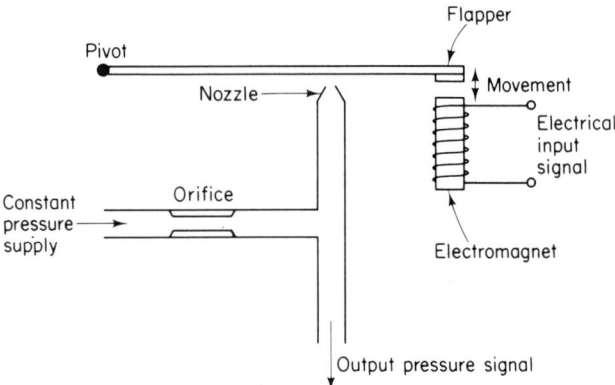

Figure 6.16   Electrical signal to pressure signal change

signal being changed from electrical to pressure change. The electrical
signal is used to energise an electromagnet which then moves a flapper
in front of the nozzle of a pneumatic comparator (see Chapter 2 and
Figure 2.8). The amount of air that can escape from the nozzle depends
on the distance of the flapper from the nozzle. If the nozzle is completely sealed off by the flapper then the output pressure rises to the
value of the supply pressure. The further away the flapper is from the
nozzle the lower the output pressure. Thus the movement of the flapper
as a result of the electrical signal received results in a related pressure
change output.

SERVOMECHANISMS

One of the main control elements in a car is the steering wheel. By use
of it the car can be made to follow the desired path along the road.
Figure 6.6 illustrated this control system. In small cars the power put
into turning the steering wheel is that power which is used to turn the
wheels of the car so that it follows the required path. In a large car the
power required to turn the wheels is more than a driver can easily supply
and thus without any other mechanism such a car is difficult to steer.
In such cases the car is provided with *power-assisted steering*. The small
amount of power used to turn the steering wheel is used to control the
much larger amount of power used to turn the wheels of the car. The
control system has *power amplification.*

A control system employing power amplification is called a *servomechanism* if it is used to control the position, velocity or acceleration
of some item. *Figure 6.17* shows one basic form of a servomechanism.

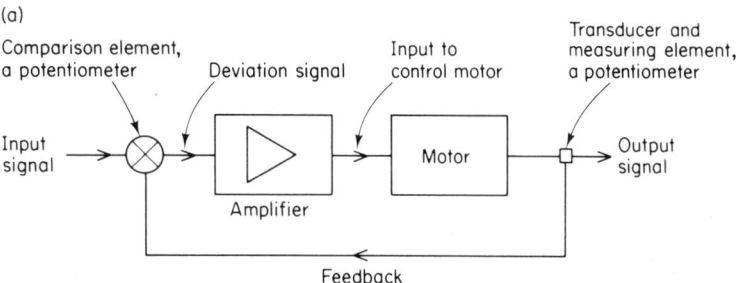

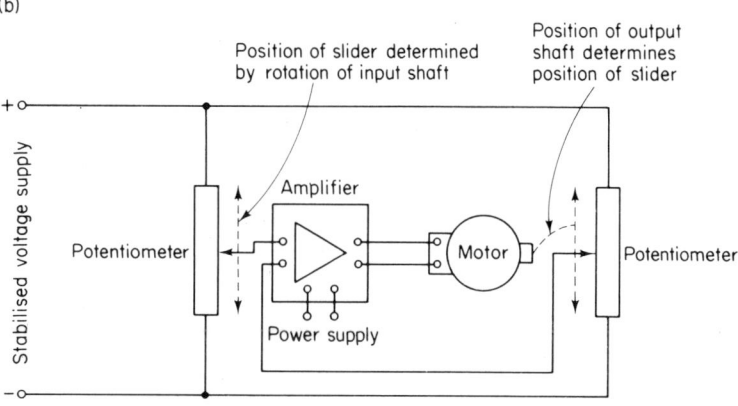

Figure 6.17   A simple servomechanism

This could be the form of the servomechanism used for the power-assisted steering or certainly the control of movement of one shaft by means of movement of another shaft. Thus the input signal could be the rotation of the steering wheel, the output the rotation of the drive shaft determining the position of the car wheels. The movement of the steering wheel results in a signal being fed into the amplifier, the steering wheel movement causing a slider to move across a potentiometer. The output from the amplifier, a power amplifier, results in a motor shaft rotating. The shaft of the motor, the output, is linked to a slider moving across a potentiometer. The movement of the slider results in a change in the potential difference fed back to the amplifier. The amplifier thus has two inputs—that from the steering wheel and that from the feedback signal. If these are the same then there is no output from the amplifier and the motor ceases to rotate. Thus the motor rotates as long as there is a difference between the input from the steering wheel and the feedback signal. When the angular positions of the steering wheel shaft and that of the motor shaft are the same there is no feedback. The system thus aligns the two shafts. If the steering wheel is then rotated the motor shaft will be made to rotate until its alignment is the same as that of the steering wheel.

Figure 4.13 showed how the servomechanism system could be used to control the movement of the pen in a recorder display unit. *Figure 3.23* showed the same principle used in the self-balancing potentiometer.

THE POTENTIOMETER

The potentiometer is used widely in control circuits, generally as a comparison element. Most potentiometers consist of many turns of wire wound on a former with a contact arm sliding over these turns and making electrical contact with them. The following are some of the terms associated with the specifications of potentiometers.

The *resolution* of a potentiometer is defined as the potential difference between adjacent turns of the wire on the former divided by the input potential difference to the potentiometer as a whole. The slider can only make contact with the wire at certain points when it is wound on a former and so the output potential difference from the potentiometer can change only in steps (*Figure 6.18*). Some forms of potentiometer use a thin strip of resistance material rather than a coil and so have virtually infinite resolution.

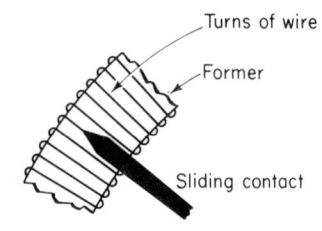

Turns of wire

Former

Sliding contact

Figure 6.18

Potentiometers can be designed so that there is a direct proportionality between the angle through which the slider has rotated and the output potential difference. Such potentiometers are said to be *linear potentiometers*. A *sine potentiometer* is designed to have an output potential difference proportional to $V \sin \theta$, where θ is the angle through which the slider has rotated. A *logarithmic potentiometer* has an output potential difference proportional to $\log \theta$.

In the case of a linear potentiometer the theoretical relationship between the output potential difference and the angle θ through which the slider has rotated is a straight line graph passing through the origin (*Figure 6.19*). This requires not only absolute uniformity of the wire but also virtually infinite resolution. *Figure 6.19* as well as showing the theoretical relationship also shows the type of variation from this that might reasonably be expected to occur in practice. The term *absolute linearity* is used to give a measure of the amount of deviation from the theoretical straight line. The absolute linearity is defined as the maximum

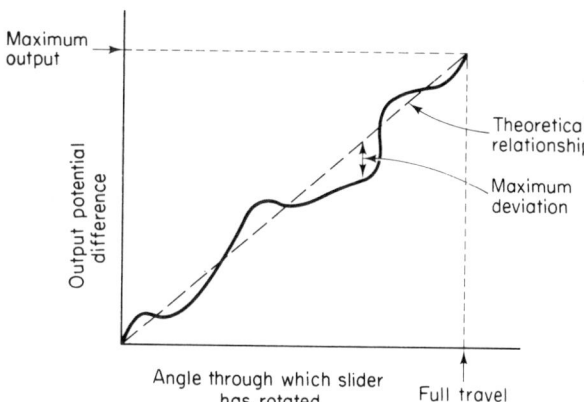

Figure 6.19

deviation from the straight line expressed as a percentage of the full
output potential difference. Thus if a potentiometer shows a maximum
deviation from the straight line of 0.050 V and the maximum output
potential difference is 10 V then the absolute linearity is

$$\frac{0.050}{10} \times 100 = 0.50\%$$

Another way of giving a measure of the variation from linearity is
to specify the *independent linearity*. This is defined in terms of the
maximum deviation from the best straight line drawn to minimise the
deviations.

For those potentiometers where the output is not linear with respect
to the angle rotated the term *conformity* is used. The *absolute conformity* is the maximum deviation from the theoretical relationship
expressed as a percentage of the total applied potential difference.

The movement of the slider over the resistance track can result in
spurious electrical signals termed *noise*. This noise or *output smoothness*
of the potentiometer is expressed as a percentage of the total applied
potential difference under specified conditions of use.

More than one arrangement of slider moving over a resistance track
can be mounted on the same slider shaft. Each of these arrangements is
known as a *cup* and an arrangement of two or more cups is called a
gang.

Table 6.1 Typical example of a technical data sheet for a linear potentiometer.

The RCP 09 is a compact 0.875 inch diameter conductive plastics track
potentiometer. It is an extremely cost effective instrument with all the performance advantages of conductive plastics. Resistance values of 5k or 1 kΩ
and independent linearity of 0.25% are available.

Performance specification

Resistance ± 10%	5000Ω or 1000Ω
Electrical angle ± 2°	342°
Independent linearity	± 1.0%, 0.5%, 0.25%
Resolution	virtually infinite
Power dissipation at 20°C	1.5 watt
Insulation at 500 V d.c.	100 MΩ minimum
Temperature range	−65°C to 130°C

(Penny and Giles Conductive Plastics Ltd, Blackwood, Gwent)

SYNCHRO SYSTEMS   The term *synchro* is used to describe a group of devices which use electrical signals for the transmission of control information over a distance. They are sometimes referred to as a.c. position motors.

Figure 6.20 shows the basic arrangement of a synchro element. Essentially it consists of a stator and a rotor. The stator is the part that does not rotate, the rotor is the part that does rotate. The stator carries three coils positioned at 120° intervals around the stator case. The wires from each coil are arranged so that one wire from each coil is connected to a

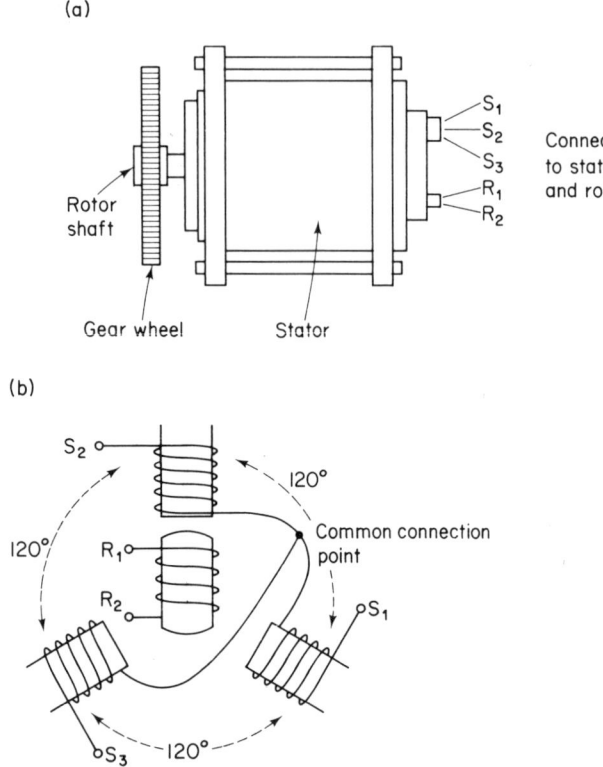

Figure 6.20 A synchro element
(a) External appearance (b) Basic wiring arrangement

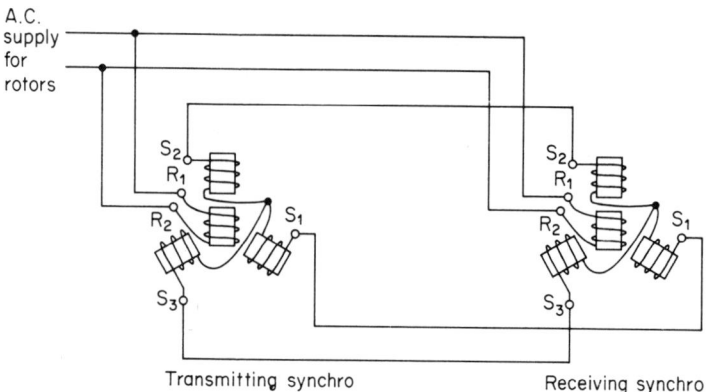

Figure 6.21 A simple synchronous transmission link

common point and the other wire forms an output lead, labelled S_1, S_2 and S_3. The rotor consists of a laminated iron core carrying a single coil wound on it. The wires from this coil are brought out via slip-rings, the output being labelled R_1 and R_2. Alternating current is supplied to the rotor resulting in e.m.f.s being induced in each of the stator coils. The e.m.f. induced in any one coil will depend on the position of the rotor relative to it.

A simple synchronous transmission link consists of one synchro element as the transmitter and one as the receiver (*Figure 6.21*). When the rotors in the two elements are in the same positions, as in the diagram, then the e.m.f.s in corresponding coils in the two stators are the same. Thus, for example, S_2 in the transmitting synchro has the same e.m.f. as S_2 in the receiving synchro. There is therefore no potential difference between the two terminals and thus no current flows. If, however, the rotors are not at the same position in each synchro element then the e.m.f.s are not the same in corresponding coils and there is a potential difference between corresponding terminals and hence a current flow. The current passes through the stator coils and produces a magnetic field which results in a force on the rotors causing them to become aligned in the same directions. Thus if the rotor of the transmitting synchro is rotated to some particular angle the rotor in the receiving synchro will be made to rotate to the same angle.

One of the applications of a synchronous link is for the remote indication of the position of a shaft, the transmitting synchro being mechanically coupled to the shaft so that the rotor of the synchro is in alignment with the shaft. The receiving synchro can be some distance away, the position of its rotor being always the same as that of the transmitting synchro and hence the shaft. The torque given by synchros, of the form described above, is only sufficient for operating indicators or very light drives.

Figure 6.22 shows how two synchros can be used in a position control system. The synchros are not being used in the same way as in *Figure 6.21* where they were used for transmitting information. There the two rotors were connected to the same alternating current supply, in the control circuit only the transmitting synchro has its rotor connected to the alternating current supply. The other synchro, referred

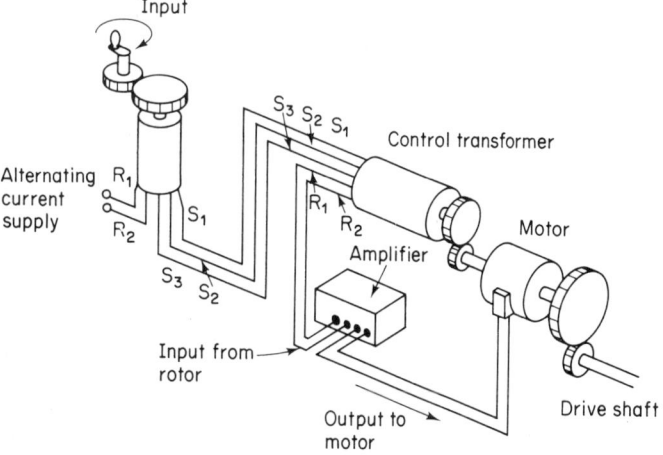

Figure 6.22 A simple synchro control of position

to as the control transformer, has its rotor connected to an amplifier.

The alternating current in the rotor of the transmitting synchro induces e.m.f.s in the stator coils. Because the stator coils of the two synchros are connected a current flows through the stator coils of the control transformer. The alternating current induces an e.m.f. in the control transformer rotor coil. The size of the induced e.m.f. depends on the positions of both rotors. The output to the amplifier from the control transformer rotor is a signal which depends on the difference in position between the two rotors. This signal is then amplified and used to operate a motor. The motor rotates the control transformer rotor shaft in such a direction as to reduce the output from that rotor to zero, i.e. it lines up the shafts of the two rotors. Thus a rotation of the shaft of the transmitter rotor results in a rotation of a drive shaft, mechanically coupled to the control transformer rotor shaft, until both the shafts are in the same alignment. This type of control system could be used by a machine operator whereby rotating a handle causes a turntable to be rotated.

PROBLEMS

(1)   Distinguish between open-loop and closed-loop control systems.

(2)   Distinguish between positive and negative feedback.

(3)   Explain the terms shown in *Figure 6.23* and the function of each of the components shown in the system.

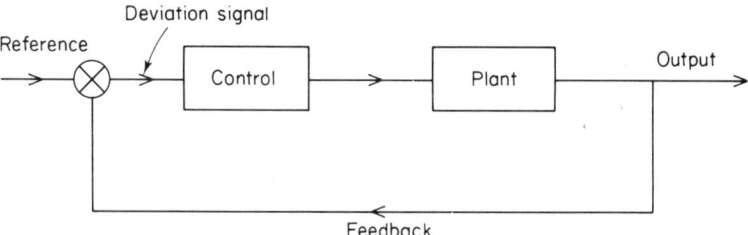

Figure 6.23   A closed loop control system

(4)   Explain how instability can occur with a control system employing negative feedback.

(5)   Explain the term process control.

(6)   *Figure 6.24* shows a number of control systems. Identify the function of each part of a system and explain how control is achieved.

(7)   Propose control systems for the following:

(a)   An automatic steering system for the control of a car moving along a specified track.

(b)   An automatic control of the level of a powder in a hopper.

(c)   An automatic system controlling the mass of a powder packed into each packet on a conveyor belt.

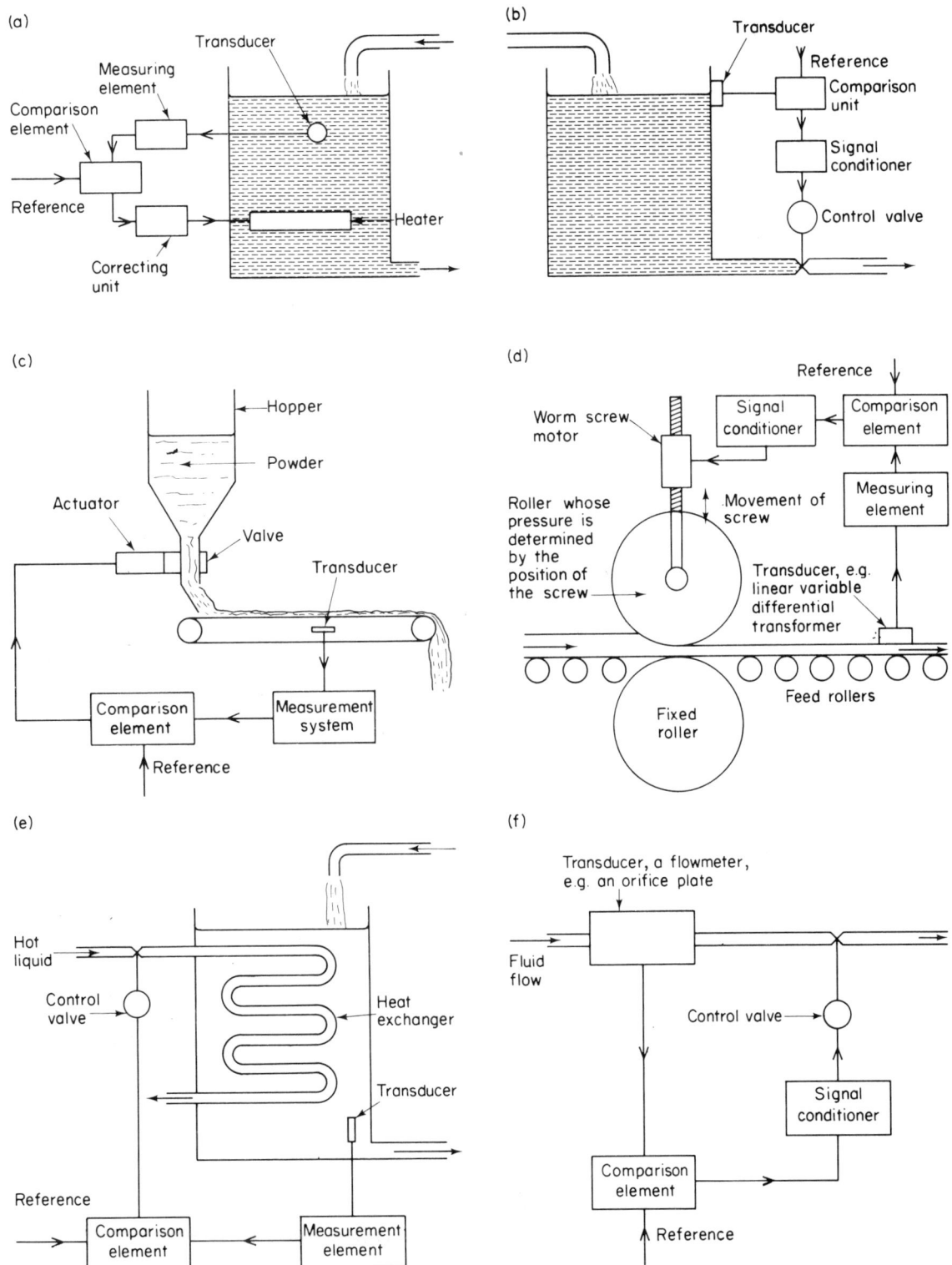

Figure 6.24 (a) An automatic temperature controller (b) An automatic level control system (c) Automatic control of the flow of a powder from a hopper (d) Automatic control of the thickness of a product (e) Automatic control of temperature in a tank (f) Automatic control of the rate of fluid flow along a pipe

(d) An automatic control system for the temperature in a furnace.

(e) An automatic control for the tension in a spring.

(f) An automatic control for the rate of flow of air as draught into a furnace.

(8) (a) Suggest possible transducers for use in the control systems shown in *Figure 6.24a, b, c, e*.

(b) In *Figure 6.24d* a linear variable differential transformer is suggested as the transducer. Explain how this transducer operates in this context.

(c) In *Figure 6.24f* an orifice plate flowmeter is suggested as the transducer. Explain how this transducer operates in this context.

(9)   The following is taken from a manufacturer's data sheet (IVO Counters Ltd, Croydon, England).

Heavy duty electromagnetic *predetermining counter* Type FS 218.

These newly developed predetermining counters have large figures over 7 mm high, ensuring easy readability. Also for ease of presetting and resetting, the pushbuttons have been set apart.

These five-figure counters have been designed for use throughout the engineering industry where processes are controlled by preset counters giving an output signal to a machine after the predetermined number of electrical impulses have been recorded. Typically, the applications include: weighing; liquid measurement; length control; continuous weighing machines; coding machines; winding machines and for dosing installations.

The FS 218 series of counters operate in a subtracting mode. Each count records a full step, i.e. each impulse subtracts one count from the preset figure set on the counter. A microswitch within the counter is operated when all figures reach zero. The counter will reset to the original preset number when either the reset button is depressed, or a remote reset impulse is received, or by means of a built-in automatic reset. Resetting the counter automatically returns the microswitch to its initial position and allows the counter to repeat its previous cycle.

(a) From a consideration of the above what do you understand by a predetermining counter?

(b) Incorporate such a counter in a design for the control of the 'doses' of liquid issued from a pipe. The pipe is required to deliver predetermined volumes of a liquid into containers. The system could be the petrol pump which can be preset by pressing a button so that it delivers a stated volume of petrol. Specify the other components used in your design.

(c) Incorporate such a counter in a design for the control of the number of items packed into a container. The same number of the object are to be packed into every container and thus a control is needed to count the objects passing along a belt and to either stop or divert the flow when the number of objects

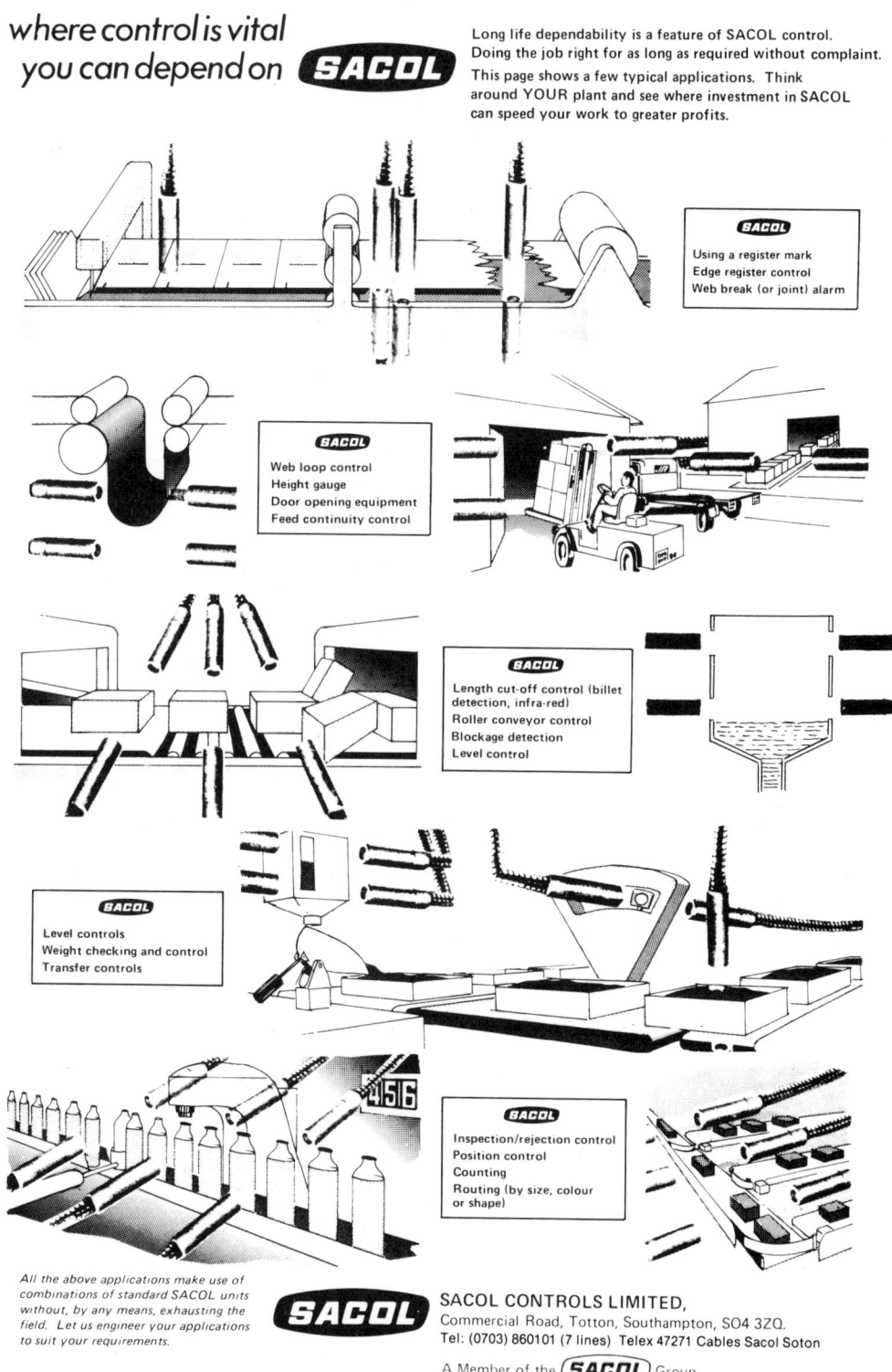

**where control is vital
you can depend on** SACOL

Long life dependability is a feature of SACOL control. Doing the job right for as long as required without complaint.

This page shows a few typical applications. Think around YOUR plant and see where investment in SACOL can speed your work to greater profits.

SACOL
Using a register mark
Edge register control
Web break (or joint) alarm

SACOL
Web loop control
Height gauge
Door opening equipment
Feed continuity control

SACOL
Length cut-off control (billet detection, infra-red)
Roller conveyor control
Blockage detection
Level control

SACOL
Level controls
Weight checking and control
Transfer controls

SACOL
Inspection/rejection control
Position control
Counting
Routing (by size, colour or shape)

All the above applications make use of combinations of standard SACOL units without, by any means, exhausting the field. Let us engineer your applications to suit your requirements.

SACOL **SACOL CONTROLS LIMITED,**
Commercial Road, Totton, Southampton, SO4 3ZQ.
Tel: (0703) 860101 (7 lines) Telex 47271 Cables Sacol Soton

A Member of the SACOL Group

Figure 6.25   (reproduced by permission of Sacol Controls Ltd)

has reached some predetermined number. Specify the other components used in your design.

(10) *Table 6.1* gives the specification of a potentiometer.
(a) What are the maximum deviations from the potential difference from a perfectly linear potentiometer that might be expected?
(b) Why is the insulation of significance for a potentiometer?
(c) Comment on the resolution.

(11) Describe and explain the action of a control valve.

(12) Distinguish between the operating characteristics of the following types of control valve: quick-acting type, linear type, equal percentage type.

(13) Explain what is meant by the term servomechanism.

(14) Design a servomechanism for the control of the rudder of a ship by the captain on the bridge.

(15) Explain how synchros can be used for (a) a transmission link, (b) a position control system.

(16) *Figure 6.25* is taken from a manufacturer's catalogue (Sacol Controls Ltd, Southampton) and illustrates the wide variety of control operations possible with photoelectric units.

Devise an automatic control for one of the operations indicated in the figure, showing the closed loop by means of a block diagram and explaining the function of each block. State what characteristics you would require of each of the components.

ASSIGNMENTS

(1) The water level control arrangement described in *Figure 6.10* uses a ball-lever system. Investigate such a system, determining the accuracy with which the level can be obtained and the general characteristics of the system. Does the system hunt? How does the system react to slow rates and to rapid rates of change in the water taken from the tank?

(2) Investigate some control system and write a report in which the system is analysed and its characteristics commented on.

(3) Monitor the room temperature in a room where the temperature is controlled by means of a thermostat. Correlate the temperature with the actions of the thermostat.

(4) Devise an automatic system for the opening of a vent when the temperature in a greenhouse or some other enclosure rises.

7 System response

Aims: At the end of this chapter you should be able to:
Explain what is meant by the free oscillation of a system and the effect of damping on such oscillations.
Describe the behaviour of systems when subject to oscillating disturbances.

FREE OSCILLATIONS   If you pull the bob of a simple pendulum to one side (*Figure 7.1a*) and then let go, the pendulum will swing backwards and forwards, i.e. an oscillation occurs. The time taken for each complete oscillation remains the same, i.e. the frequency of the oscillation is constant, even though the amplitude of the oscillation gradually diminishes. *Figure 7.1b* shows

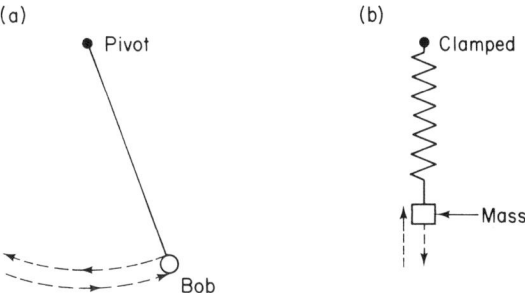

Figure 7.1 (a) Simple pendulum (b) Spring-mass system

a mass attached to the lower end of a vertical spring, the upper end of the spring being clamped. When the mass is pulled down and then released an oscillation occurs. Both these oscillating bodies, the pendulum and the spring, are producing what are called *free* or *natural oscillations.*

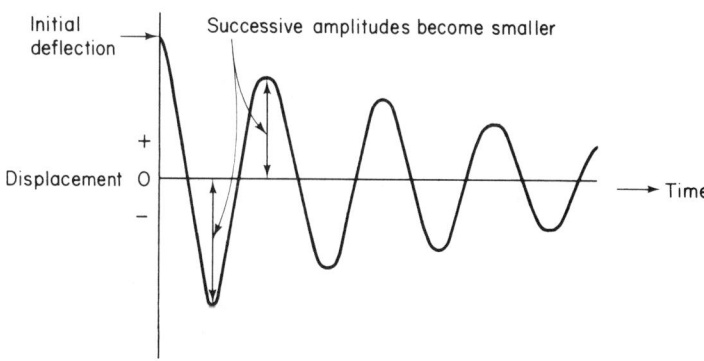

Figure 7.2 Damped oscillations

These oscillations occur when no force is applied to sustain the oscillations once they have started. There are other forces acting on the oscillating objects, e.g. gravity and damping forces, but they are not applied to sustain the oscillations.

All systems are in some ways like the spring-mass system. All systems are capable of free oscillations. In many cases the oscillations are rapidly damped out, i.e. they soon die away. *Figure 7.2* shows the form of damped oscillations likely to occur with the spring-mass system. The amplitude of the oscillation steadily diminishes with time.

The pointer of an instrument, e.g. a galvanometer, will oscillate when given an impulse, equivalent to pulling it to one side and then allowing it to swing. The system has a certain natural frequency at which it oscillates. *Figure 7.3* shows the type of oscillation that might occur. The pointer does not swing instantaneously to its final reading but overshoots and then oscillates around that value for a while. The pointer takes time to settle down to its final reading.

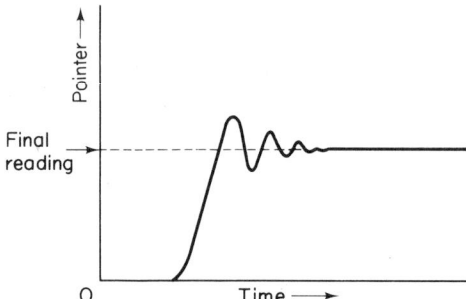

Figure 7.3 Oscillations of an instrument pointer

If you press down on the bumper of a car and then let go, the body of the car will oscillate up and down. The motion is damped and so the oscillation quickly dies away. The system has a certain natural frequency.

Buildings have natural frequencies at which they will oscillate when given an initial push. A sharp gust of wind can set a building into oscillation just like the spring-mass system was set into oscillation by an initial pull to one side. Machines have natural frequencies at which they oscillate when given the initial push or pull and then allowed to freely oscillate. All systems are capable of free oscillations.

DAMPING   The free vibrations of physical systems all die away with time. A mass oscillating on the end of a spring as a result of an initial impulse will in each successive oscillation decrease in the amplitude of its motion, eventually coming to rest. The motion is said to be *damped*. Damping is a factor that is always present whether it is wanted or not. In the design of measuring systems damping is often deliberately introduced. It would be very inconvenient if the pointer of a galvanometer kept on oscillating for some considerable time after the current had been supplied to the instrument. Obtaining the reading would be very difficult. The motion of the galvanometer pointer is thus generally deliberately damped so that the pointer reaches the value that is to be read in a reasonable amount of time.

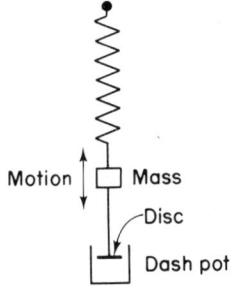

Figure 7.4 Dashpot damping

Figure 7.4 shows a simple form of damping applied to the spring-mass system. The damping arrangement is known as a *dashpot*. It consists of a disc attached to the end of the oscillating spring, the disc then oscillating inside an enclosure, the pot, which has a cross-sectional area just slightly greater than that of the disc. The pot may contain only air though often it is filled with a light oil. The effect of the dashpot arrangement on the motion of the spring-mass oscillator depends on the fluid in the pot, the size of the disc and the size of the gap between the disc and the pot walls.

Figure 7.5 shows the different types of result that can be achieved with different degrees of damping when the spring-mass system is subject to an initial impulse. In the ideal unreal case the oscillations

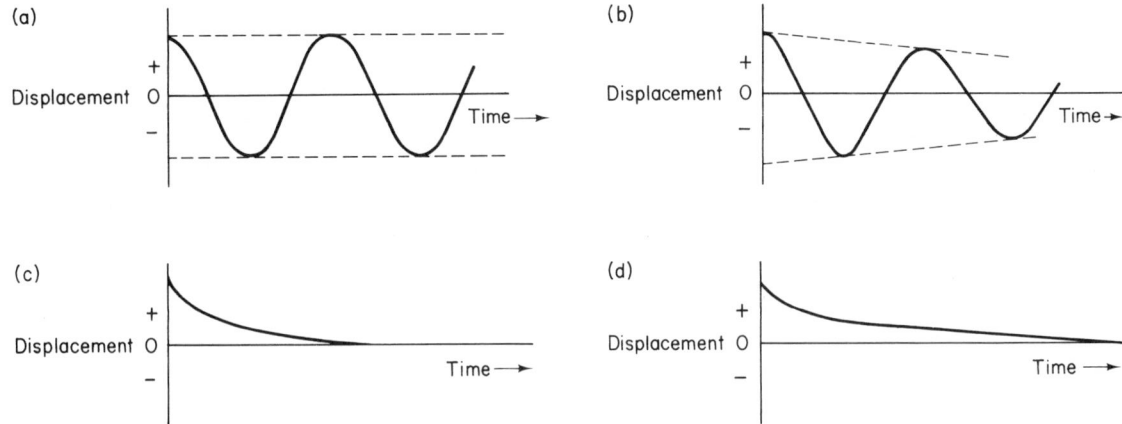

Figure 7.5 (a) No damping, an ideal unreal situation (b) Lightly damped (c) Critically damped (d) Overdamped

would continue on for ever with undiminished amplitude. With light damping the amplitude of the oscillation continually diminishes but only gradually. As the damping is increased so the number of oscillations that occur becomes less until at a particular damping, known as the *critical damping*, the oscillator just fails to oscillate. The displacement

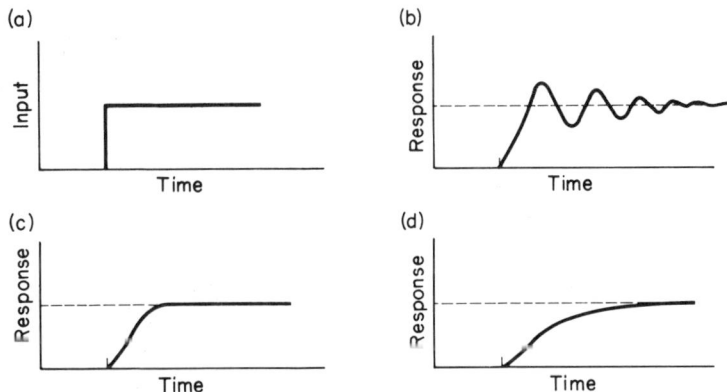

Figure 7.6 (a) A step input (b) Lightly damped response (c) Critically damped response (d) Overdamped response

of the oscillator at this critical damping gradually declines to zero. The oscillator is overdamped if the damping is in excess of that needed to give the critically damped condition.

An instrument such as a galvanometer which is overdamped is said to be *dead-beat.* An instrument is said to be *aperiodic* when the motion is either critically damped or overdamped.

The input to many systems, particularly galvanometers, is usually a step input (*Figure 7.6a*). This is an input which suddenly rises in magnitude and levels off at a constant value. *Figure 7.6b,c,d* shows how the response of the system to such an input depends on the degree of damping.

RESONANCE

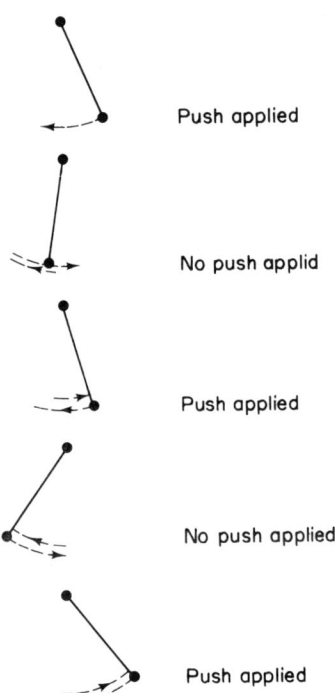

Push applied

No push applid

Push applied

No push applied

Push applied

Figure 7.7 Building up the oscillation of a pendulum by pushing at just the right moments

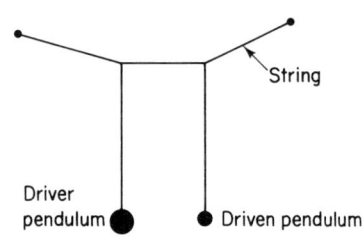

String

Driver pendulum

Driven pendulum

Figure 7.8

A freely oscillating pendulum has an amplitude which slowly decreases with time. Eventually the oscillation ceases completely. The pendulum is started in operation by being pulled to one side and then released, i.e. a force is applied initially to cause the motion and then no longer applied. If however the force is repeated at regular intervals, say every time the pendulum reaches one extreme of its swing, then the oscillation can be maintained with not only an undiminished amplitude but an increasing amplitude (*Figure 7.7*). The force has to be repeated at regular intervals and at the right moments. The situation is comparable to that of a child on a swing being pushed by a parent. The parent has to keep giving pushes at just the right moments for the amplitude of the swing's motion to build up. The important thing is for the pushes to be at precisely the right moments. Try pushing a child on a swing if you do *not* push at the right moments! The pushes have to be in phase with the swing's motion. If one oscillation is completed in, say, 2 s, i.e. a frequency of $\frac{1}{2}$ Hz, then the pushes have to be applied every 2 s, i.e. with the same frequency. Large amplitude oscillations of the pendulum or the swing can only be built up if the applied frequency of the pushes is the same as the free oscillation frequency of the system.

Figure 7.8 shows another way in which forces can be applied to a pendulum. The forces are applied through the pendulum's support by another pendulum, the driver pendulum. The force being applied varies sinusoidally with time. When the lengths of the driver and driven pendulums are the same the driving frequency and the driven frequency are the same and large amplitude oscillations build up. If the driver pendulum does not have the same length the amplitude of the oscillations of the driven pendulum are much smaller. When the frequency of the driver pendulum is the same as that of the driven pendulum and a large amplitude is built up then the effect is called *resonance.*

This effect is not restricted to pendulums. When the frequency applied to a system equals the free oscillation frequency of the system then resonance occurs. *Figure 7.9* shows how the displacement of a system varies with time for different applied frequencies. The large amplitude build-up occurs only when the applied frequency is equal to the free oscillation frequency of the system.

The amplitude to which any oscillating system builds up depends on not only how the applied frequency relates to the free oscillation frequency of the system but also on the damping of the system. The less the damping the greater the build up of amplitude (*Figure 7.10*).

A dashpot system can be used to increase the damping of a system and so reduce the effects of oscillations close to the free oscillation frequencies of the system. *Figure 7.11* shows a possible arrangement

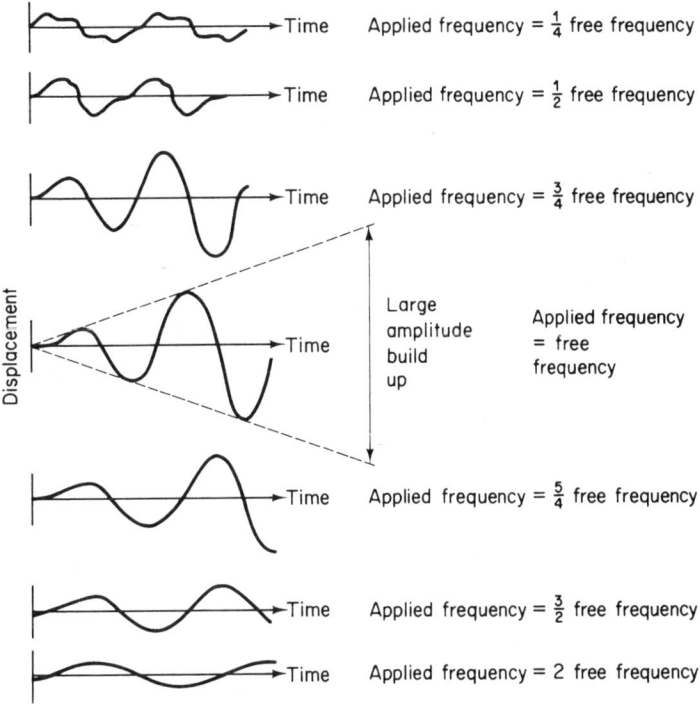

Figure 7.9

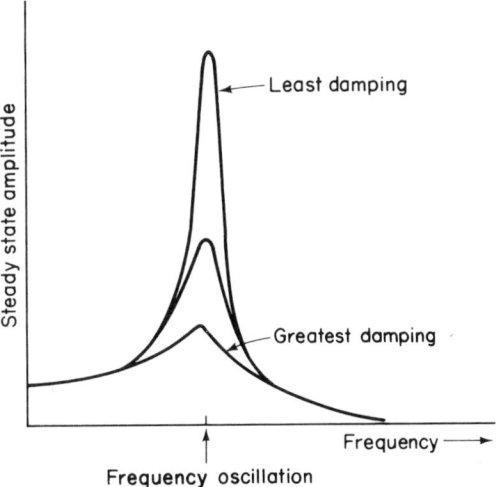

Figure 7.10 The effect of damping on the amplitude of a system subject to an applied frequency

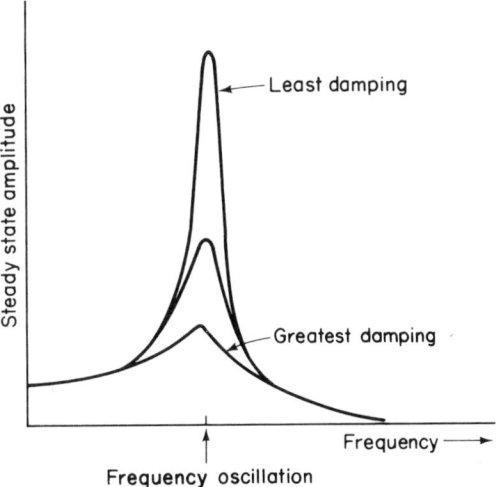

Figure 7.11 An anti-vibration mounting

that could be used as a flexible mounting for some object with mass. Such an arrangement is known as an *anti-vibration mounting*.

A rubber pad can be considered as being equivalent to a spring and dashpot combined, it having the property of being able to oscillate like a spring but also having internal damping. Thus in some cases a rubber pad might be an appropriate mounting to reduce the effect of external oscillations on some mass. The seat cushion in a car or on a train has the same function—to reduce for the passenger the amplitude of the oscillations arising from the motion.

Springs are often used as the means of mounting objects so that they are not significantly affected by vibrations. The spring will oscillate but is chosen to have a free oscillation frequency far removed from the frequencies likely to be encountered. The car suspension spring is chosen to have a low free oscillation frequency, all but the very low frequencies are then 'blocked' by the spring. The car suspension with such springs is said to be 'soft'.

Instruments have to be protected from vibrations if they are to give readings which are steady and consistent. Thus in the case of an accurate weighing machine an anti-vibration mounting may have to be used if the full accuracy of the instrument is to be realised.

PROBLEMS

(1)   State what is meant by the terms 'free oscillation frequency' and 'resonance'.

(2)   Describe the behaviour of overdamped, critically damped and underdamped systems when subject to a step input.

(3)   What factors determine the amplitude with which a system will oscillate when subject to a force which varies sinusoidally with time?

(4)   A moving-coil galvanometer has a dead-beat movement. How would you expect the indicator of that instrument to behave when the meter is abruptly switched into a circuit carrying a current?

ASSIGNMENTS

(1)   Examine the mounting of a machine and comment on the form of mounting used and the efficiency of that mounting in preventing the communication of vibration both to and from the machine.

(2)   Investigate the factors determining the degree of damping produced by a dashpot. Consider the size of the disc, the size of the gap between the disc and the pot walls, and the oil used.

Appendix A
Units

The units used in this book are in the 'Système International d'Unités' or SI system. These units are based on metric units and though now widely used throughout the world other units will be found in some countries and also in references to older measuring systems. The main other system that might be encountered is known as the foot-pound-second or FPS system. The following table compares the units in the SI and FPS systems.

Quantity	*FPS unit*		*SI unit*
Length	1 yard (yd)	=	0.914 m
	1 foot (ft)	=	304.8 mm
	1 inch (in)	=	25.4 mm
Area	1 square yard	=	0.8361 m^2
	1 square foot	=	92.90×10^3 mm^2
	1 square inch	=	645.2 mm^2
Volume	1 cubic yard	=	0.7646 m^3
	1 cubic foot	=	0.028 317 m^3
	1 cubic inch	=	16.387×10^3 mm^3
	1 gallon (gal)	=	4.5461×10^{-3} m^3
	1 pint (pt)	=	0.5683×10^{-3} m^3
Mass	1 pound (lb)	=	0.4536 kg
	1 ton	=	1.016 tonne = 1.016×10^3 kg
	1 slug	=	14.594 kg
Density	1 lb/ft^3	=	16.019 kg/m^3
	1 lb/in^3	=	27.680×10^3 kg/m^3
	1 slug/ft^3	=	0.5154×10^3 kg/m^3
Velocity	1 ft/s	=	0.3048 m/s
Force	1 poundal(pdl)	=	0.1383 N
	1 lbf	=	0.4536 kgf = 4.4482 N
	1 tonf	=	1.0161×10^3 kgf = 9.964 kN
Pressure, stress	1 lbf/in^2 (p.s.i.)	=	0.07031×10^{-4} kgf/m^2 = 6.895 kN/m^2
	1 lbf/ft^2	=	4.882 kgf/m^2 = 0.047 88 kN/m^2
	1 tonf/in^2	=	0.1575×10^{-4} kgf/m^2 = 15.44 MN/m^2
Temperature	1 in Hg	=	25.4 mm Hg = 3.386 kN/m^2
	t°F	=	$5(t - 32)/9$ $^\circ$C

Appendix B
Report writing

The assignments in many instances require the writing of reports. The following article reproduced from *Engineering*, February 1978, should help in the production of readable reports.

Ten ways to make your reports and publications unreadable

Most of us get involved in writing for others, be it project reports, conference papers, or sales literature and it is very easy to make these offerings almost useless to the recipient. Some of the potential pitfalls are outlined by Ian Parker of Industrial Handbooks Ltd

1 Enjoy the product. Be full of what you and your team have done, and use the document to demonstrate your own cleverness without concerning yourself with what the eventual reader wants to know.

2 Stick to jargon. Use the 'short-hand' that you use with your immediate colleagues; this should ensure that it will be largely meaningless to the reader, while being perfectly clear to you and your people.

3 Don't ask for professional help; even though you are not a professional writer, you are undoubtedly perfectly capable of coping with the most complex descriptive or instructional text.

4 Make your points in the order of their importance to you; no need to confuse yourself with an introduction to set the scene, or even a contents list so people know what to expect. Leave your secretary to lay the document out; she's the typist so she should have the final decision on what is stressed by the layout, how headlines are treated, etc.

5 Never bother to check whether there are company, national, or even international standards to which your document should be written – it will take up a few minutes of your time to check, whereas otherwise someone else can re-write it for you.

6 Don't draw conclusions, or justify them; just spell out the facts – interspersed with a few prejudices of course – and leave the reader to sort himself out as best he can.

7 Develop an impressive writing-style, with long flowing sentences – every one should include a set of dashes – while brackets too are a must (preferably with several pairs (perhaps one pair within the other for absolute clarity)).

8 Demonstrate the depth of your education with the length of your words; never use a short word when you can find a long one. This will be particularly impressive to such audiences as an arab sheikh reading a sales pitch, or a factory hand trying to repair a machine.

9 Ensure the document is as long as possible; inclusion of any test results you can find, all the available drawings, etc., will impress your audience with how hard you have been working.

10 Don't leave yourself too much time to write the document; we all of us write better under pressure, and if you leave yourself the opportunity for second thoughts and corrections you will lose the original sparkle of your prose.

Remember above all that the important things are that you should enjoy writing the document, and that it shouldn't interrupt the smooth running of your day. Your readers (your superiors, your firm's customers . . .) will then be kept on their toes trying to find out what they need to know.